# The Science of Science Communication III

Inspiring Novel Collaborations and Building Capacity

PROCEEDINGS OF A COLLOQUIUM

Steve Olson, *Rapporteur*

Held on November 16-17, 2017,
at the National Academy of Sciences in Washington, DC

NATIONAL ACADEMY OF SCIENCES

THE NATIONAL ACADEMIES PRESS
*Washington, DC*
www.nap.edu

THE NATIONAL ACADEMIES PRESS   500 Fifth Street, NW   Washington, DC 20001

This publication is based on the Arthur M. Sackler Colloquium of the National Academy of Sciences, "The Science of Science Communication III," held November 16–17, 2017, at the National Academy of Sciences building on Constitution Avenue in Washington, DC.

Support for the Arthur M. Sackler Colloquia is provided by the Dame Jillian & Dr. Arthur M. Sackler Foundation for the Arts, Sciences, and Humanities. Additional support for this colloquium was provided by the Alfred P. Sloan Foundation, The Annenberg Public Policy Center of the University of Pennsylvania, the Burroughs Wellcome Fund, the Gordon and Betty Moore Foundation, The Kavli Foundation, the Penn State Science Communication Program, the Rita Allen Foundation, Science Sandbox—a Simons Foundation initiative, and the William and Flora Hewlett Foundation. Any opinions, finding, conclusions, or recommendations do not necessarily reflect the views of any organization or agency that provided support for the project.

International Standard Book Number-13:   978-0-309-46858-9
International Standard Book Number-10:   0-309-46858-2
Digital Object Identifier:   https://doi.org/10.17226/24958

Additional copies of this publication are available from the National Academies Press, 500 Fifth Street, NW, Keck 360, Washington DC 20001; (800) 624-6242 or (202) 334-3313; http://www.nap.edu.

Cover art: The Duke & the Duck

Copyright 2018 by the National Academy of Sciences. All rights reserved.

Printed in the United States of America

Suggested citation: NAS (National Academy of Sciences). *The Science of Science Communication III: Inspiring Novel Collaborations and Building Capacity: Proceedings of a Colloquium.* Washington, DC: The National Academies Press. doi: https://doi.org/10.17226/24958.

*The National Academies of*
# SCIENCES · ENGINEERING · MEDICINE

The **National Academy of Sciences** was established in 1863 by an Act of Congress, signed by President Lincoln, as a private, nongovernmental institution to advise the nation on issues related to science and technology. Members are elected by their peers for outstanding contributions to research. Dr. Marcia McNutt is president.

The **National Academy of Engineering** was established in 1964 under the charter of the National Academy of Sciences to bring the practices of engineering to advising the nation. Members are elected by their peers for extraordinary contributions to engineering. Dr. C. D. Mote, Jr., is president.

The **National Academy of Medicine** (formerly the Institute of Medicine) was established in 1970 under the charter of the National Academy of Sciences to advise the nation on medical and health issues. Members are elected by their peers for distinguished contributions to medicine and health. Dr. Victor J. Dzau is president.

The three Academies work together as the **National Academies of Sciences, Engineering, and Medicine** to provide independent, objective analysis and advice to the nation and conduct other activities to solve complex problems and inform public policy decisions. The National Academies also encourage education and research, recognize outstanding contributions to knowledge, and increase public understanding in matters of science, engineering, and medicine.

Learn more about the National Academies of Sciences, Engineering, and Medicine at **www.nationalacademies.org**.

*The National Academies of*
SCIENCES • ENGINEERING • MEDICINE

**Consensus Study Reports** published by the National Academies of Sciences, Engineering, and Medicine document the evidence-based consensus on the study's statement of task by an authoring committee of experts. Reports typically include findings, conclusions, and recommendations based on information gathered by the committee and the committee's deliberations. Each report has been subjected to a rigorous and independent peer-review process and it represents the position of the National Academies on the statement of task.

**Proceedings** published by the National Academies of Sciences, Engineering, and Medicine chronicle the presentations and discussions at a workshop, symposium, or other event convened by the National Academies. The statements and opinions contained in proceedings are those of the participants and are not endorsed by other participants, the planning committee, or the National Academies.

For information about other products and activities of the National Academies, please visit www.nationalacademies.org/about/whatwedo.

# Contents

PREFACE     vii

1    *COMMUNICATING SCIENCE EFFECTIVELY: A RESEARCH AGENDA*     1
Building the Capacity for Research on Science Communication, 4
Communicating Uncertainty, 6

2    A LIFE AT THE INTERSECTION OF SCIENCE AND SOCIETY     7
Distinguishing Between Science and Pseudoscience, 9
Truth Seeking Through Dialogue, 10

3    SCIENCE COMMUNICATION IN A POLITICALLY CHARGED ENVIRONMENT     13
Political Polarization and Science Communication, 14
Communicating About Immigration, 15
Communicating About Climate Change, 17
Countering Vaccine Hesitancy, 19
When Does Science Misinformation Matter?, 22
The Influence of Science, Health, and Cultural Literacy, 23

4    CREATING A COLLABORATIVE COMMUNITY     25
Managing Conflicts in Science Communication, 26
Bridging Boundaries in Science Communication, 27
Collaborative Science Communication Research, 29

| 5 | INCENTIVES IN SCIENCE COMMUNICATION | 31 |
|---|---|---|
| | Factors That Influence Faculty Behavior, 33 | |
| | Ambassadors for Science and Engineering, 34 | |
| | Revising Tenure Guidelines, 36 | |
| | Melding Art with Research, 36 | |
| | The Science Behind the News: Human Gene Editing, 37 | |
| 6 | COMMUNICATING WITH POLICY MAKERS | 41 |
| | Congratulations and Cautions, 44 | |
| 7 | THREATS TO SCIENCE'S REPUTATION | 45 |
| | The "Science Is Broken" Narrative, 46 | |
| | Enhancing Trust in Science, 47 | |
| | Addressing Detrimental Practices in Science, 48 | |
| | Promoting Transparency and Openness, 50 | |
| | The Science Behind the News: Driverless Cars, 52 | |
| 8 | EVALUATING SCIENCE COMMUNICATION | 55 |
| | Influence on Twitter, 56 | |
| | Scientists on Twitter, 57 | |
| | Communicating Big Ideas in the Social Sciences, 58 | |
| 9 | COMMUNICATING UNCERTAINTY | 61 |
| | Predictions Without Measures of Uncertainty, 62 | |
| | The Risks of Ignoring Uncertainty, 63 | |
| | The Science Behind the News: Gene Drive, 63 | |
| 10 | THE ROLE OF PHILANTHROPY IN SCIENCE COMMUNICATION | 67 |
| | Promoting Science Communication Among Scientists, 68 | |
| | Science Communication for Philanthropists, 69 | |
| 11 | REFLECTIONS ON THE COLLOQUIUM | 71 |
| | Stories, Rewards, and Relationships, 72 | |
| | The Power of Stories, 73 | |

REFERENCES 75

APPENDIXES

| A | Agenda | 79 |
|---|---|---|
| B | Speakers | 85 |

# Preface

Climate intervention ... fracking ... vaccines ... human genome editing ... artificial intelligence.... With so many complex, important, and sometimes uncertain scientific issues facing our society, there has never been a more critical time to communicate science effectively. Polls show that the public has high trust and confidence in science and believes that science provides great benefits to the world. Yet, many people continue to deny the conclusions of science in areas such as evolution, climate change, and vaccination. Why do people refuse to accept the scientific consensus on these topics given their high confidence in science generally?

The science of science communication can help resolve this riddle. Behavioral and cognitive scientists can explore how people perceive and think about scientific issues. Social scientists can study the contexts in which science communication occurs and the effects of that communication on public policy. These interdependent research areas have revealed much about how people interpret and sometimes reject scientific information, with much more to be learned.

In 2011, the late Ralph J. Cicerone, then President of the National Academy of Sciences (NAS) and a noted climate scientist, had the vision to convene a diverse group of scientists and practitioners to survey the state of research on science communication and its application in practice. His initiative led to the first Arthur M. Sackler Colloquium on the Science of Science Communication, which was hosted by the NAS in Washington, DC, in 2012. It brought together leading social, behavioral, and decision scientists to survey the state of the art in their fields as they relate to science communication along with prominent communication practitioners and policy makers, including four science advisors to the president of the United States. A special issue of the *Proceedings of the National Academy of Sciences (PNAS)*[1] made the research available to a wide audience, as did the webcast and archived videos[2] of the sessions.

The second Arthur M. Sackler Colloquium on the Science of Science Communication was held the following year. It expanded the set of contributing sciences and highlighted the particular challenges of communicating about contested or controversial science and of establishing working relationships that take full advantage of the sciences of communication. It, too, led to a special issue of *PNAS*[3] and archived video,[4] as well as a published summary.[5] Together, the first and second colloquia were a major impetus behind the 2017 consensus study *Communicating Science Effectively: A Research Agenda* from the National Academies of Sciences, Engineering, and Medicine.[6]

The third Arthur M. Sackler Colloquium on the Science of Science Communication, held on November 16–17, 2017, in Washington, DC, used *Communicating Science Effectively* as a framework for examining how one might apply its lessons to research and practice. It considered opportunities for creating and applying the science along with the barriers to doing so, such as the incentive systems in academic institutions and the perils of communicating science in polarized environments. Special attention was given to the organization and infrastructure necessary for building capacity in science communication. More than 550 people attended the colloquium; the webcast had more than 16,000 live views. Archived video[7] is available, with another special issue of *PNAS* forthcoming.

The sponsors of the colloquium were the Dame Jillian & Dr. Arthur M. Sackler Foundation for the Arts, Sciences, and Humanities, the Alfred P. Sloan Foundation, The Annenberg Public Policy Center of the University of Pennsylvania, the Burroughs Wellcome Fund, the Gordon and Betty Moore Foundation, The Kavli Foundation, the Penn State Science Communication Program, the Rita Allen Foundation, Science Sandbox—a Simons Foundation initiative, and the William and Flora Hewlett Foundation. The colloquium owed a special debt to Dame Jillian Sackler, whose gift

---

[1] See http://www.pnas.org/content/110/supplement_3.
[2] See http://www.nasonline.org/programs/sackler-colloquia/completed_colloquia/agenda-science-communication.html.
[3] See http://www.pnas.org/content/111/supplement_4.
[4] See http://www.nasonline.org/programs/sackler-collquia/completed_colloquia/agenda-science-communication-II.html.
[5] See https://www.nap.edu/catalog/18478.
[6] See https://www.nap.edu/catalog/23674.
[7] See https://www.youtube.com/user/sacklercolloquia/videos?disable_polymer=1.

in memory of her husband, Arthur M. Sackler, has made possible four or five colloquia each year that bring together leading scientists in their fields with others interested in the topics of their research. More information about the colloquia is available online.[8]

At the National Academies, Marty Perreault and Susan Marty helped make the colloquium happen. Frank Sesno and Ashley Llorens moderated the first and second days of the event. Steve Olson wrote the summary of the colloquium, with editorial and administrative assistance from Stephen Mautner.

Finally, we want to especially thank Barbara Kline Pope, executive director for communications and the National Academies Press at the National Academies (until September 2017 and now director of the Johns Hopkins University Press) for her leadership on all three of the Arthur M. Sackler Colloquia on the Science of Science Communication.

Karen S. Cook
Baruch Fischhoff
Alan I. Leshner
Dietram A. Scheufele
*Colloquium Organizers*

Marcia McNutt
*President*, National Academy of Sciences

---

[8] See http://www.nasonline.org/programs/sackler-colloquia.

# 1

# *Communicating Science Effectively: A Research Agenda*

---

> **Important Points Made by the Presenters**
> 
> - Communicating science effectively requires aligning strategy with goals, tailoring messages to audiences, dealing with complexity and uncertainty, going beyond the "deficit model" of science communication, and framing issues so that audiences will be receptive to them. (Leshner)
> - Building the capacity to do research on science communication requires organizations to have knowledge of the science relevant to communication, awareness of the need to employ that science, proper staffing, internal and external coordination, and incentives. (Fischhoff)
> - The incentives needed for collaborative, multidisciplinary research include rewards to scientists both for publishing in top journals and for demonstrating the usefulness of their work. (Fischhoff)

The need to communicate the results of scientific research to the public has never been greater, but effective science communication is complex and must be learned, said Alan Leshner, chief executive officer emeritus of the American Association for the Advancement of Science, in the first session of the third Arthur M. Sackler Colloquium on the Science of Science Communication. Today's media environment is fragmented and polarized. Communicators increasingly use social media that have fewer gatekeepers to establish facts. The news cycle is fast paced, with many voices competing for attention.

Leshner chaired the committee that produced the 2017 report *Communicating Science Effectively: A Research Agenda*, which served as a framework for the colloquium (NASEM, 2017a). He began by laying out five cross-cutting themes from the report.

The first involves the need to align strategy with goals. These goals may be sharing the findings and excitement of science, increasing appreciation for science, increasing knowledge of a specific issue, influencing opinions or behaviors, or considering public perspectives and finding common ground. The key, said Leshner, is to establish goals and develop strategies to accomplish those goals.

The second theme is the need to tailor messages to specific audiences, including the general public, scientists, media, advocacy groups, corporations, nonprofit research organizations, health professionals, government agencies, science enthusiasts, policy makers, political commentators, and individual activists. "This is the biggest mistake that I see people make," said Leshner. "They talk about what they want to talk about, not what people want to hear."

The third theme is that communicating science is particularly difficult because scientific information is complex and often uncertain. Scientists may take uncertainty as a fact of life, Leshner observed, but members of the public can react to it by saying, "If you're not really sure what you're talking about, then maybe I can do whatever I want."

The fourth theme is that the deficit model of science communication is, for the most part, wrong. This model holds that people would make choices more consistent with the scientific evidence if they were provided with more information through "better" science communication. But people rarely make decisions based only on scientific information, Leshner pointed out. "People draw on their own beliefs about the world. They use their own analogies, metaphors, and prior experiences." Scientists lose their credibility with other scientists if they ignore science, but members of the public can disregard, deny, or distort findings with few immediate consequences.

The fifth and final theme Leshner cited is that the framing of an issue is critical. Science communicators cannot simply state facts the way they want to state them. They need to consider how to frame those facts so that specific audiences will be receptive to the issues they embody.

Leshner urged science communicators to "go glocal." Communicators need to take a global issue and make it meaningful at the local or the personal level, he said. He also emphasized the necessity of public engagement, which involves considering not only style and content but the intent of a conversation. Converting a monologue to a dialogue requires communicating *with* the public rather than *to* the public. Most scientists are not prepared to listen to and respect public concerns, but they can learn to do it. "Engagement is an acquired skill, as is communication more broadly," he said.

Leshner concluded by highlighting some of the research questions posed by the report. With regard to the complexities of communicating science, a major question is

- What are the important individual and social factors that shape the effectiveness of science communication for different audiences, and how do they interact in various contexts?

In the area of engaging formally with the public, important research questions include

- What structures and processes for public engagement are most effective?
- To what degree do these approaches need to be tailored to the diversity of the participants, the decisions to be made, and the nature of the topic?

In the special case of policy-maker audiences, important research questions are

- How is scientific information used by policy makers in formal policy processes?
- How can science communication affect policy processes?
- Does it matter who the communicator is?

When public controversy adds to the complexity, researchers need to ask

- How can science be communicated effectively amid conflicts over beliefs and values?
- What are effective ways of communicating scientific consensus, as well as degrees or types of uncertainty?

Finally, in a complex and competitive media environment, questions to answer include

- How can competing messages and sources of information be better understood?

- How can social media platforms, online games, and blogs be used?
- Are some forms of media better than others for achieving certain science communication goals?

Building the science communication enterprise will require researcher-practitioner partnerships, interdisciplinary work and dialogue, recruiting more researchers from neighboring disciplines, randomized controlled trials, and the use of big data, Leshner said. As more becomes known about what works in science communication, communicators need to know about and use the results of this research. "Let me end with a request," Leshner concluded. "Let's see if we can have science replace intuition and common sense about what works in science communication."

## BUILDING THE CAPACITY FOR RESEARCH ON SCIENCE COMMUNICATION

According to Baruch Fischhoff, Howard Heinz University Professor in the Institute for Politics and Strategy and the Department of Engineering and Public Policy at Carnegie Mellon University, building the capacity to do the kind of research called for in *Communicating Science Effectively* requires five components:

- Science,
- Awareness,
- Staffing,
- Coordination, and
- Incentives.

The two previous Arthur M. Sackler Colloquia on the Science of Science Communication (Fischhoff, 2013; NAS, 2014) highlighted the science developed to date, as have other reports from the National Academies of Sciences, Engineering, and Medicine, Fischhoff said. *Improving Risk Communication* (NRC, 1989) built on the work of formal and informal science educators to point to different and more effective ways of communicating. *Toward Environmental Justice: Research, Education, and Health Policy Needs* (IOM, 1999) called for two-way communication with the populations concerned with environmental justice. The workshop proceedings *Building Communication Capacity to Counter Infectious Disease Threats* (NASEM, 2017b) examined the capacity to communicate effectively about infectious diseases. Another workshop summary, *Potential Risks and Benefits of Gain-of-Function Research* (IOM and NRC, 2015), looked at the communications required to build awareness of pathogens with pandemic potential. The

U.S. Food and Drug Administration (FDA) has sought to characterize and communicate uncertainty in developing the benefit-risk frameworks it employs, as in the workshop summary *Characterizing and Communicating Uncertainty in the Assessment of Benefits and Risks of Pharmaceutical Products* (IOM, 2014). *Intelligence Analysis for Tomorrow: Advances from the Behavioral and Social Sciences* (NRC, 2011), which was written by a committee led by Fischhoff, included guidance on how to communicate intelligence analysis.

The second capacity-building requirement, awareness of the need for communication science, would seem self-evident, but it is not, Fischhoff observed. People overestimate how well they understand others, and vice versa. As a result, they unwittingly communicate poorly and then blame their audiences. "Somehow or other, we need to swim upstream against the assumption that people don't need this research," he said. To heighten awareness of this need, the report of the Risk Communication Advisory Committee that Fischhoff chaired for FDA (Fischhoff et al., 2011) briefly summarized the science, pointed out the implications of that science, and showed how to evaluate science communication for no money at all, for a little money, or for an amount of money commensurate with the personal, organizational, and political stakes riding on effective communication.

The third necessity for research on science communication is proper staffing with people who have the full suite of requisite expertise. That means subject-matter specialists who can provide accuracy, decision scientists who can ensure relevance, behavioral scientists who can enhance comprehensibility, and practitioners who can guide execution. All opinions are welcome, said Fischhoff, because anyone can have an insight on a topic, but authority needs to be invested in those who know the most. "You don't want your psychologists rewriting things for greater comprehensibility and getting the science wrong," he said. "And you don't want scientists taking a communication opportunity as a teachable moment to talk about their own research."

Research capacity requires coordination to allow for collaboration among the people who have these forms of expertise, Fischhoff continued. As an example of internal coordination, he pointed to a project led by Diana Rhoten and Denise Caruso at the Hybrid Vigor Institute that studied multidisciplinary centers to see how well they were working (Rhoten, 2003). That project used network analysis to study the transfer of information within centers. Similarly, as an example of external coordination, FDA has studied the communication principles involved in communicating risk (Fischhoff, 2015).

The final capacity-building requirement is to provide incentives for people to stay the course in collaborative, multidisciplinary research. Incentives for scientists must include rewards for both publishing in top journals and demonstrating the usefulness of their research. Fischhoff said that he has had the good fortune throughout his career to work at institutions that have had incentives for scientists to apply their findings to practical problems. This entails both applied basic science, which evaluates accepted science in applied contexts, and basic applied science, which pursues fundamental topics arising in applied contexts. In either case, basic science and applied science "need one another," he said.

## COMMUNICATING UNCERTAINTY

During the question-and-answer session, Leshner and Fischhoff elaborated on two key points: the role of uncertainty in science communication, and the incentives that exist in science to communicate research results.

The public tends not to hear the caveats and clauses in science communication, observed Leshner. When he was director of the National Institute on Drug Abuse, he often described addiction as "a brain disease expressed in behavioral ways and in a social context." But "nobody ever heard the clauses," he pointed out. "All they ever heard was that addiction is a brain disease. You have to find a way to couch or explain what you're talking about in a way that they can grasp."

Fischhoff observed that people would love to have certainty if it exists, "but we live with uncertainty all the time, and the question is how do you present it in a way that people feel is more honest." Researchers have studied how to express uncertainty in ways that people will understand, which provides a foundation for applications. If communicators promise but cannot deliver certainty, people will feel betrayed. "You need to test those messages, because they're not immediately obvious," he said.

Spurred by a question from the audience, both Leshner and Fischhoff pointed to the lack of incentives within the scientific community to engage in science communication. But Leshner expressed the opinion that the situation is slowly changing. "Young scientists get it," he said. "They do understand, and they want to do it. . . . As graduate education evolves, we will, more and more, see young scientists educated in how to effectively communicate, and it will roll into the incentive system."

Fischhoff was somewhat less optimistic. "It would take a brave junior person to risk their promotion and tenure process to devote time to this," he said. Instead, the incentive structure needs to change within science to enable all scientists to contribute to the commons of public understanding and good will. At the same time, all scientists need access to the research on how to best communicate their science.

# 2

# A Life at the Intersection of Science and Society

---

> **Important Points Made by the Presenter**
> - Public mistrust of science has been increasing over time. (Gawande)
> - Distinguishing science from pseudoscience, providing explanatory narratives based on science, and exposing the tactics used to mislead people can all help members of the public understand science. (Gawande)
> - Discussing and pursuing ideas with curiosity, inquisitiveness, openness, and discipline can model the truth seeking that is a hallmark of science. (Gawande)

Atul Gawande, the keynote speaker at the colloquium, said that he has played three different roles at the intersection of science and society. He is a scientist who has gathered and analyzed data in public health experiments designed to reduce deaths and improve the quality of life. He is a staff writer with *The New Yorker* and the author of four books on critical issues in medicine and health. He is also a surgeon at Brigham and Women's Hospital, where he has opportunities to put into practice the ideas developed through his research and writing.

In each of these three roles, he has encountered public mistrust of science. Research findings that suggest changes in medical practice often encounter resistance from physicians and patients, he said. He has "learned not to read the comments section" of his articles, in part to avoid the charge that he is disseminating "fake news." Patients of his have even raised doubts about the trustworthiness of his plans as a surgeon. He asked, what accounts for this mistrust?

Part of the answer involves a misunderstanding of science. As Gawande defined it, science requires a commitment to a way of building knowledge and explaining nature through factual observation and testing. This "is not a normal way of thinking," he said. "Much of what we do in science may be counterintuitive. A scientific explanation will not necessarily be the same as the explanation that may come from the wisdom of divinity or the explanations that come from experience or common sense. We watch the sun move across the sky. It's common sense. It's moving across the sky. Or people get colds in cold weather, and cold must produce colds."

In science, intuitions are hypotheses that need to be tested. He particularly likes Edwin Hubble's description of a scientist as someone with "a healthy skepticism, suspended judgment, and disciplined imagination." Gawande's interpretation of this description is that scientists have an experimental mind, not a litigious mind. Scientists are not free of opinion, but evidence contradicting their explanations can always arise. Hubble also said that "the scientist explains the world by successive approximations." This approach to understanding has proved remarkably powerful over time.

Yet, scientists are not necessarily trusted, despite their openness to experience and evidence, Gawande pointed out. Many people believe that childhood vaccines cause autism, even though a massive evidence base indicates that they do not. People believe they are safer if they own a gun, but they are not. People believe that climate change is not happening or is unrelated to human activities, even though it is.

Once embedded in people's minds, beliefs that are not in accord with evidence can be hard to change. This is especially the case when people do not trust scientific authorities, and trust in science has been declining among some groups over time. Distressingly, surveys done since the 1970s have demonstrated that the higher a person's educational level, the more substantial is the average decline in trust in the scientific community, particularly among conservatives. In 1974, educated conservatives had the highest level of trust in the scientific community; now they are the group with the lowest level of trust in the scientific community.

People do not dismiss scientific authority, Gawande pointed out. "As a society, the belief in the power and the fruits of science is strong." Rather, people dismiss scientific authorities and point to alternative bodies of information. This makes the job of science communication particularly crucial.

## DISTINGUISHING BETWEEN SCIENCE AND PSEUDOSCIENCE

A major task for science communicators is to distinguish between science and pseudoscience, Gawande pointed out. He cited five hallmarks of pseudoscience identified by writer Mark Huffnagle. First, people pushing pseudoscience allege a conspiracy to suppress dissenting views. Second, they provide fake experts with contrary views but no credible scientific track record. Third, they cherry-pick data and papers to discredit a field. Fourth, they deploy false analogies and logical fallacies. Fifth, they set expectations for research that are impossible to meet.

Pseudoscience takes the form of science without the substance of science. For that reason, using science to rebut typically does not work and can even backfire by strengthening the conviction of believers. "It spreads familiarity with the belief," said Gawande. "That's partly because of the nature of the brain. Misinformation sticks." Debunking a commonsense view of the world can also leave a painful gap, which causes people to raise their defenses to avoid that discomfort.

Rebutting bad science may not be effective, but asserting scientific facts is, Gawande observed. Mental models based on misinformation need to be replaced with explanatory narratives based on science, he said. With vaccines, for example, focusing on vaccine myths is far less effective than focusing on the fact that giving children vaccines has proved vastly safer than not giving them vaccines. When federal policy reduced access to funds for vaccination among the poor, vaccination fell to the point that the United States had an outbreak between 1989 and 1991 of 55,000 measles cases and 123 deaths. "It's safe to give vaccines and it's deadly not to," said Gawande, which is the message people need to receive.

Another effective approach is to expose the tactics that are used to mislead people. Bad science has a pattern, and helping people recognize that pattern helps them identify bad science when they see it. "Having a scientific understanding of the world is about helping people understand how you judge which information to trust." But the task is complicated by the appearance of new data and the development of new understandings. Bisphenol A (BPA), for example, is a carbon-based compound that has been widely used in plastics. A confluence of studies has suggested that the compound may have negative effects on human health, but the issue has been disputed. How can people decide what to believe? The science surrounding these kinds of issues has become "too vast and too complex" for most people to master. Scientists themselves can become "bullheaded, enamored of pet theories, dismissive of new evidence, and heedless of their fallibility," Gawande warned. The consensus may also be wrong, as was the case when, for years, physicians assured their patients that they were very unlikely to become addicted to pain killers by taking them after surgery.

The key in deciding what to think, said Gawande, is to consider the difference between science and pseudoscience. Does one of the sides cherry-pick data to support its view? Does one of the sides honestly grapple with evidence that runs counter to its views? Does one of the sides "assess the totality of the views versus taking the litigious position?"

The virtues of a scientific orientation lie more in the community and the body of work than in any one individual, Gawande explained. Science is a social enterprise with an intricate division of cognitive labor. "As a community endeavor, it's been beautifully, amazingly self-correcting over time. It is not beautifully organized, however. Up close, it can be an extremely rickety-looking vehicle for arriving at truth," he said. The peer-review process can be muddled, journal articles badly written, and scientific pronouncements pompous. "And it will not get any prettier," he added. "That's a contradictory idea for us to embrace: elitism in our method, but a democratization of data." Yet, that is how science works.

## TRUTH SEEKING THROUGH DIALOGUE

The increasing number of people who question scientific conclusions is adding to the inherent messiness of the scientific process, said Gawande. His surgical practice is a good example. Science is "arguably the most powerful collective enterprise in human history," yet many of his patients are deeply skeptical about even the most basic scientific knowledge. "In my work as a surgeon, there is fundamental skepticism on a regular basis about the most basic knowledge from 'mainstream science,'" said Gawande (adding, however, that there is only one kind of science). Furthermore, the ones with the greatest doubts are often the most educated.

In dealing with such patients, it is a mistake, Gawande said, to think that his scientific credentials give him any special authority. "What a scientific education offers is something that's more important: understanding what real truth seeking looks like." In his clinic visits, Gawande seeks to engage in a dialogue that reflects the process of seeking truth. "Discussing and pursuing ideas with curiosity, inquisitiveness, openness, and discipline is the way I have to get through my clinic." When a patient has a curable cancer but refuses to undergo surgery, he asks the patient to watch the progression of the cancer with him. "I've never gone through all the way to watching [a cancer] metastasize and kill somebody. Eventually, as we watch our way through that and grapple with the evidence, we have made that change."

Science communicators need to take a similar approach, he said. They need to engage with ideas and their audiences using curiosity, inquisitiveness, openness, and discipline. As an example, Gawande cited his work on the issue of how to educate clinicians to talk to patients as they face the end of life. Role playing enables clinicians to go beyond just providing facts to patients and then asking them to decide what to do. Rather, clinicians can learn to take the role of a counselor who can elicit people's goals.

> What's your understanding of where you are with your illness at this time? What would you like to know about what might be ahead for you from me? What would you like to know about your prognosis from me? What are your fears for your future with your health? What are your goals if your health worsens? What are you willing to go through? What are you not willing to go through for the sake of more time? What's the minimum quality of life you find acceptable?

Asking these questions makes it less frightening for both patients and clinicians to talk about uncertainties, establish goals, and then try to meet those goals.

Another useful approach is to focus on positive rather than negative outcomes. Solutions journalism, for example, stresses competence and why taking certain steps can lead to improvements. Such stories can introduce people to puzzles or threats to identify crucial questions that need to be answered. For example, rather than writing about the threats and quandaries posed by end-of-life questions, writing about people who work in palliative care and geriatrics can emphasize the skills needed to talk through difficult decisions and arrive at the best possible decisions. "People with solutions and directions for solutions for major issues in our lives, major threats to our world, that is something people can attach themselves to and believe in," Gawande said.

All science communicators can use these kinds of techniques to learn how to communicate better, Gawande concluded, and they need to do so. "The stakes for understanding this could not be any higher than they are today, because we're not just battling anymore what it means to be scientists. We're battling for what it means to be citizens."

# 3

# Science Communication in a Politically Charged Environment

---

> **Important Points Made by the Presenters**
>
> - The polarization of viewpoints in America has profoundly changed how people receive and respond to scientific information. (Iyengar)
> - Social media and Internet news sources have become major sources of information for many people. (Massey)
> - Communication is more effective if it addresses what people need and want to know. (Bruine de Bruin; Morgan)
> - Though many communication strategies are under way in the United States, they often are underfunded and not evaluated. (Finley; Nyhan)
> - The socioeconomic circumstances people face may be more important than science communication in affecting vaccination rates. (Hornik)
> - Language, culture, and education all create variation in what people hear. (Scrimshaw)

The political climate for science communication has changed even in the 4 years since the second Arthur M. Sackler Colloquium on the Science of Science Communication. People are more likely to get scientific information from social media or politically partisan news sources. Political polarization has made some people less likely to trust information from particular news outlets. Scientists are increasingly likely to communicate directly with lay audiences through social media, sometimes before peer review of their research.

Several speakers at the colloquium explored the challenges facing science communication in a politically charged environment, including the challenges that arise in communicating information about immigration, climate change, and vaccination.

## POLITICAL POLARIZATION AND SCIENCE COMMUNICATION

Americans' sense of political identity has resulted in an increasingly powerful form of ingroup-outgroup polarization, said Shanto Iyengar, Chandler Chair in Communication at Stanford University. Surveys by the American National Election Studies, The Pew Charitable Trusts, and other organizations document increasing hostility toward the outparty, along with increased social distance depending on party affiliation. Other results suggest that implicit bias against the outparty exceeds the levels of implicit bias based on race (Iyengar and Westwood, 2015). Behavioral measures of interpersonal trust demonstrate that the party cleavage is deeper than social cleavages even in divided societies.

Social identity theory holds that humans tend to gravitate to groups and that once they form a group they develop hostile feelings toward outgroups, Iyengar pointed out. In the political domain, researchers have coined the term *sorting* to indicate that the political divide between Democrats and Republicans is now being amplified through reinforcing cleavages. This sorting—where a partisan divide converges with racial, gender, age-based, religious, and urban–rural divides—contributes to a situation in which partisan opponents are considered what Iyengar termed "the repugnant cultural other." As an example of this effect, disapproval of interparty marriage has grown dramatically in the United States, though not in the United Kingdom, since the 1960s (Iyengar et al., 2012). Partisanship has become a litmus test for interpersonal relationships, including those based on family, kinship, friendship, and online networks. As a result, "groups are increasingly politically homogeneous," Iyengar pointed out.

This polarization has had a profound effect on media in the United States. Many news organizations cater to partisan preferences. Partisans, especially on the right, impute bias to mainstream news—for example, a recent poll found that 46 percent of respondents think that the media make up stories about President Trump. Partisan news sources have seen increased circulation, creating the potential for an "echo chamber," where people mostly receive information that reinforces their preferences.

When partisans encounter information at odds with their identity, they counterargue and often move further away from the evidence-based positions, said Iyengar. The credibility of universities, think tanks, and government agencies is considered suspect. When a political party has developed a reputation or a stance on a particular policy issue, the adherents of that party tend to follow the cues supplied by party leaders.

Provision of expert information can change the preferences of policy makers, Iyengar observed, though sometimes in counterintuitive ways. For example, when researchers presented information to policy makers showing that needle exchange programs do not increase drug use, liberal policy makers' support for these programs remained close to 100 percent before and after the treatment. But conservative policy makers tended to disagree with the programs *even more* after they received the information. "If anything, the suggestion from this study is that expertise can have a boomerang effect," he said.

Iyengar concluded by proposing several ways of dampening the polarization. Gentler rhetoric from high places would help, but "I would not be losing a lot of sleep over that possibility." Increased interpersonal contact across the party divide could reduce animus; for example, in marriages that mix Democrats and Republicans, evaluations of the 2016 presidential candidates tended to be less extreme. Increasing the mobilization of nonpartisans would be helpful but not easy. Finally, with topics such as immigration, Iyengar pointed out that framing the issue in terms of individuals rather than policy can increase public support for a position.

## COMMUNICATING ABOUT IMMIGRATION

Doug Massey, Henry G. Bryant Professor of Sociology and Public Affairs at the Woodrow Wilson School of Public and International Affairs at Princeton University, has been engaged with public communication on immigration issues since he got his Ph.D. in 1978 from Princeton. He has penned op-ed articles, written and edited books for general readers, appeared on radio and television programs, and testified in front of the U.S. Congress. "I've gotten better at communicating," he said. "But in my experience, it's not the communication that has become the problem. It's the context into which you're trying to communicate, which has changed dramatically."

Massey served on the committee that produced the report *The Integration of Immigrants into American Society* (NASEM, 2015) and he was a reviewer of the report *The Economic and Fiscal Consequences of Immigration* (NASEM, 2017d). The messages of both reports were well received by the mainstream press, he said. The headline for the latter report in *The New York Times* was "Immigrants Aren't Taking Americans' Jobs, New Study Finds." The headline in *The Wall Street Journal* was "Immigration Does More Good than Harm to Economy, Study Finds." However, other organizations cast the report's findings in a very different light. The restrictionist Center for Immigration Studies described the report under the headline "National Academy of Sciences Study of Immigration: Workers and Taxpayers Lose, Businesses Benefit." The Heritage Foundation's headline was "National Academy of Sciences Report Indicates Amnesty for Unlawful Immigrants Would Cost Trillions of Dollars." The *Washington Examiner* wrote "Immigration costs at least 278 billion, 'illegals' mentioned only three times." *Breitbart*'s headline was "National Academies' Study Shows $500 Billion Immigration Tax on Working Americans." *Infowars* said, "Report Explains Financial Cost of Illegal Immigrants for American Taxpayers," adding that the "wall could save taxpayers $64 million over the next decade."

These diametrically opposed interpretations point to a major structural shift in the media and in America, Massey said. Before the U.S. Congress deregulated the cable industry in 1984, there were five major sources of broadcast news: ABC, NBC, CBS, PBS, and CNN. After deregulation, these five were joined by Fox in 1986, CNBC in 1989, Bloomberg in 1990, FSTV in 1995, MSNBC in 1996, Blaze in 2011, One America in 2013, and CBSN and Newsmax in 2014. Before 1987, the Fairness Doctrine required broadcasters to devote some of their airtime to discussing controversial matters of public interest and to present contrasting views regarding those matters, and controversial issues had to be presented in a manner that was, in the view of the Federal Communications Commission, honest, equitable, and balanced. In 1987 the commission eliminated the doctrine.

Following the spread of the Internet in the 1990s, social media became a major new source of information and Internet news sites became alternatives to the mainstream press. Internet trolls began to sow discord, disbelief, and discontent on the Internet by starting arguments or upsetting people, by posting inflammatory, extraneous, or off-topic messages in online communities, and by otherwise trying to undermine facts and knowledge. In 2010 the U.S. Supreme Court ruled that freedom of speech prohibited the government from restricting independent political expenditures by profit or nonprofit corporations, which unleashed a wave of dark money into politics.

"Unfortunately, science and support for science has become a partisan issue," concluded Massey. While trust in science has remained largely unchanged among liberals and moderates since about 1980, it has declined markedly for conservatives. The result has been an increase in obfuscation and misdirection. Communicators therefore need to use all the tools available to them to support their messages, he insisted. "It's not enough simply to communicate clearly and effectively. You have to have a counterinsurgency tactic ready to fight the inevitable information wars."

## COMMUNICATING ABOUT CLIMATE CHANGE

A long-term collaboration between M. Granger Morgan, Hamerschlag University Professor of Engineering at Carnegie Mellon University, and Wändi Bruine de Bruin, university leadership chair in behavioral decision making at the Leeds University Business School, has produced several key insights and lessons learned in communicating about climate change research.

In one project, Morgan and Bruine de Bruin studied messages about removing carbon dioxide from power plant emissions and putting it deep underground, thereby helping to curb climate change by preventing carbon dioxide from entering the atmosphere. A series of interviews in the early 2000s found that most interviewees had not heard of carbon capture and sequestration (CCS) and were, at best, lukewarm about the technology after they heard a brief description of it. Instead, they tended to want to discuss and compare alternative strategies for low-carbon electricity generation, such as solar power and wind power (Palmgren et al., 2004). Participants expressed a lower degree of acceptance for geological and oceanic CCS than for alternative low-carbon strategies, including nuclear power. "What might have happened in that survey is that people focused on the negative information about CCS more than the positive information," said Bruine de Bruin. "We know from decision research that negative information gets more attention than positive information and losses loom larger than gains. What we should have been doing is give people information about the risks, costs, and benefits about all technologies so that people could recognize that all technologies have downsides and upsides."

In follow-up studies with graduate student Lauren Fleishman Mayer, the investigators wrote multiple-attribute descriptions of 10 low-carbon technologies, with the descriptions pilot tested with the intended audience to ensure that they were understandable (Fleishman et al., 2010). The descriptions covered how the technology works, carbon dioxide releases, cost, and safety. After studying the materials at home, participants preferred coal plants with CCS versus coal plants without CCS, particularly with the newer types of plants. "What we concluded from the project is that, if you give people positive and negative information about a range of technologies, they may be willing to accept the technology that you're trying to communicate about," Bruine de Bruin observed.

Based on these and other studies, Morgan, Bruine de Bruin, and their co-workers have been developing various communication materials on climate change for lay audiences. However, Morgan has increasingly come to the conclusion that, to be informed participants in public discourse about climate change issues, people need to know just three things. First, burning fossil fuel produces carbon dioxide. Second, higher levels of carbon dioxide in the atmosphere warm the planet and cause the climate to change. Third, once carbon dioxide gets into the atmosphere, much of it stays there for hundreds of years.

Studies that Morgan did with Ann Bostrom at the University of Washington had shown that progress has been made in increasing public understanding of the first two points. However, understanding of the third point was making much less headway. Therefore, in a collaboration with Ph.D. student Rachel Dryden, Morgan, Bruine de Bruin, and Bostrom set out to learn what people thought about how long carbon dioxide stays in the atmosphere. In a survey, they asked people how long they thought it would take for common air pollutants such as smog and for carbon dioxide to return to preindustrial levels if humans quit emitting them. The results showed that people did not understand the difference between the two—even though common air pollutants would leave the atmosphere quickly while carbon dioxide would stay there for hundreds of years (Dryden et al., 2017). "We find that disturbing," said Morgan, "because if you believe that, then people would be able to think: 'I don't know if this climate change stuff is real or not. But if it ever gets serious enough, we'll just fix it by reducing emissions in the same way that we fixed air pollution in places like Pittsburgh and Los Angeles.' And of course that doesn't work. Once the stuff is there, we're stuck with it." Current work is looking at communications that can help people understand why different pollutants have very different residence times in the atmosphere.

Morgan and Bruine de Bruin have drawn several conclusions from their study of climate change communication. First, communication is more effective if it addresses what people need and want to know, not just if it covers what experts think is important. Second, communications need to be worded in an understandable way so that people can use the information in their decisions. Third, it is important to test the effectiveness of communications before they are disseminated. "If you don't test your communications, then you run the risk of spending time and effort in disseminating something that doesn't help people make more informed decisions," said Bruine de Bruin.

More broadly, Morgan and Bruine de Bruin identified a need for research that brings together domain experts who can provide technical expertise and social scientists who can provide expertise on ensuring understanding with the intended audience. Interdisciplinary collaborations can pose challenges, they acknowledged, but these challenges can also be seen as strengths. Interdisciplinary researchers may aim to solve different problems, but their work together can be more useful as a result. Theories and methods may differ, but the resulting work ends up being grounded in multiple fields. Researchers may differ in academic language and culture, but they can learn to understand one another. Interdisciplinary papers may not fit into disciplinary journals, but enough top journals want interdisciplinary projects to make publication possible. Similarly, traditional academic departments may be confused by interdisciplinary research, but the departments that welcome such researchers can break new academic ground. Morgan and Bruine de Bruin have had adequate funding, supportive environments, recognition in their fields, and excellent students who have helped them balance their approaches. They have also have been fortunate to work at institutions that promote interdisciplinary work.

## COUNTERING VACCINE HESITANCY

"Vaccines work," stated Christine Finley, immunization program manager at the Vermont Department of Health. The incidence of rubella, mumps, hepatitis A, hepatitis B, measles, chicken pox, diphtheria, whooping cough, and polio—which together used to plague millions of Americans, mostly children—has dramatically decreased as vaccinations for these diseases have become widely available. Today, both polio and rubella no longer exist in the United States, and only 37 cases of polio worldwide were reported in 2016. Diphtheria, which once caused 150,000 deaths per year, is now extremely rare.

Despite these dramatic results, one in three parents in the United States today has some concern about vaccinating their children. This unease has a variety of sources, said Finley. One is that perceived benefit matters. Most parents have never seen a child with measles or who has been paralyzed by polio. However, many parents have encountered someone with autism and have heard that autism may be related to vaccines. Many other forces also come into play. Imposed risks feel scarier than risks that are chosen. Misinformation is widely disseminated through a variety of venues, especially social media. Distrust in pharmaceutical companies and in government causes people to rely on other sources of information. Many parents today believe that natural is better and fixate on the manmade nature of vaccines, even though the reaction to a vaccine is a natural response of a functioning immune system.

Vaccines are mandated for enrollment in child care and kindergarten in all 50 states, but 47 states still provide loopholes to escape these requirements. Only three states do not allow exemptions for philosophical or religious reasons—California, Mississippi, and West Virginia—with the exemptions being eliminated by California's legislature in 2015 after a measles outbreak at a Disney theme park.

"Parents' need for education and guidance in order to vaccinate with confidence is higher than ever, but we don't know exactly what that should look like," observed Finley. Nevertheless, a few things are known that work, she added. It is important to understand the audience and not to confront them in a judgmental way. It also is important to focus on the fence sitters. "People who are completely opposed to vaccines will never be interested, but they're usually less than 1 percent." The credibility of health care providers can be leveraged. "They're our partners in this, and they're very well trusted among patients." Finally, personal stories and testimonials can be very powerful.

Though many communication strategies are under way in the United States, they are often underfunded and not evaluated, Finley observed. She and Brendan Nyhan, professor in the Department of Government at Dartmouth College, have therefore formed a partnership to evaluate new approaches to promoting vaccination by determining the most effective messages and strategies in countering vaccine hesitancy. Their prospective study was one of two described at the colloquium that were chosen through a competitive process for support by the Rita Allen Foundation and the Arthur M. Sackler Colloquia.

"We should distrust our intuitions about what works and instead test them rigorously," said Nyhan. He has led two studies designed to evaluate the effectiveness of giving corrective information to people about vaccines, and both have had counterintuitive results (Nyhan and Reifler, 2015; Nyhan et al., 2014). In randomized controlled trials with nationally representative samples, the studies found that exposure to corrective information caused people to express less intention to vaccinate rather than more. Negative responses were concentrated among the people with the least favorable attitudes toward vaccines—precisely the group of greatest concern.

Finley and Nyhan's project aims to fill gaps in knowledge by bringing together scholars and practitioners who can exchange and combine insights and expertise. Practitioners have the unique perspective of engaging directly with human behavior. Academics, by partnering with practitioners, can analyze actual behavioral measures on vaccination rates rather than relying on surveys. Practitioners can also provide valuable local insights into communication problems that academics working at a broader level might overlook.

Despite the expertise that practitioners bring to the partnership, many are not trained in experimental and survey research methods. Academics can design, execute, and analyze research conducted in health care settings and apply the rigorous techniques needed to produce concrete evidence. They may be able to spot limitations in the measurement capabilities of practitioners and program officers, and they can contribute funding support to the partnership through institutional backing and grants. Nyhan pointed out that new programs can be randomized across a variety of factors. "Instead of haphazardly rolling it out, let's randomize how we roll it out, and let's partner with someone to evaluate it and see if it works."

These kinds of partnerships could become much more common, Nyhan observed. Scholars and practitioners are both interested in collaborating, "but helping them find each other, connecting people who have common interests and expertise, is harder than it might seem." Research and practice can inform each other, such partnerships are not expensive, and they can be incorporated into existing programs and activities. "What's most exciting about this partnership is the ability to have real-world impact," said Nyhan. "We're going to provide the most rigorous evidence to date of the effectiveness of vaccine messaging, which is a critically important social issue."

Doron Weber, vice president of programs and program director at the Alfred P. Sloan Foundation and one of three discussants for Finley and Nyhan's presentation, pointed out that the mindsets of hesitant parents can be changed through direct or indirect means. Monetary incentives can directly affect actions, beliefs, or behaviors. Indirectly, films, plays, books, television shows, and social media can influence the culture and the decisions people make within that culture. For example, the Sloan Foundation's support of the book *Hidden Figures*, which led to the movie of the same name about African-American female mathematicians working for NASA, could very well do as much to encourage women and underrepresented minorities to pursue science, technology, engineering, and mathematics (STEM) careers as more directly focused efforts, Weber argued. Similarly, the foundation has supported a play on vaccination and projects on the public health heroes Jonas Salk and D. A. Henderson. These cultural influences "are as powerful a weapon as we have in our arsenal, and we should not forget them."

Suzanne Ffolkes, vice president of communications at Research!America, pointed to the critical importance of the messenger and trustworthiness. Surveys have shown that Americans trust scientists and medical professionals, and the nature of this trust could be probed in research collaborations. This is one of many ways, she added, in which social and behavioral scientists could make major contributions through collaborations between practitioners and scholars. Finally, she called attention to the effects of cultural differences on vaccine hesitancy—for example, some population groups may be more likely to worry about possible side effects of vaccines than others.

Paul Hanle, president and chief executive officer of Climate Central, pointed to the parallels between the approach Finley and Nyhan are taking and the approach needed to understand public perceptions of climate change. Studies of vaccine hesitancy could yield "lessons that we might be able to take forward to apply, for example, to the extraordinarily entrenched views that are in the climate conversation." Responses may differ among groups and locations, and such differences can provide "an interesting opportunity to look at something that is more localized."

Finally, Greg Boustead of the Simons Foundation emphasized the importance of gaining greater understanding of the power of storytelling, particularly among people who might be skeptical on an issue. Science Sandbox is a strategic portfolio for grantmaking that has targeted people who do not identify as science enthusiasts, "and there is a lot we don't know and understand about what resonates with these communities." Partnerships between social and behavioral scientists and content creators can help produce the kinds of outcome measures and evaluations that funders are seeking.

## WHEN DOES SCIENCE MISINFORMATION MATTER?

Robert Hornik, Wilbur Schramm Professor of Communication and Health Policy at the University of Pennsylvania's Annenberg School of Communication, offered another perspective on the prevalence of vaccination in the United States. People have made the claim that vaccine rates have fallen because of false claims about a link between vaccine and autism, he noted. However, the percentage of 19- to 35-month-olds in the United States who have completed the seven-vaccine series rose from 2009 to 2011 and stayed constant, at about 70 percent, from 2011 to 2015. Similarly, the vaccination rate for measles, mumps, and rubella (MMR) stayed constant, at about 90 percent, from 2000 to 2016.

Few studies have specifically tested the effects of beliefs about autism on vaccination levels. Smith et al. (2008) found that selective nonreceipt of the MMR vaccine rose from 0.75 percent in 1995 to 2.00 percent in 2000 after the first autism-MMR study appeared in the medical literature, but it returned to baseline before any substantial media coverage of the issue, suggesting that the decline in vaccination rate was linked to professional wariness and not to public concerns. Other studies have assessed the presence of antivaccine content on traditional and social media, have examined beliefs about vaccine risks versus vaccine benefits, and have done cross-sectional associations of beliefs with exemption requests or, rarely, with vaccination rates. The association may be relevant in some regions or for some vaccine-resistant subgroups. But, together, the studies provide "quite limited support for the specific hypothesis that the false beliefs about vaccination and autism drive low vaccination rates."

Hornik called attention to a different association—between vaccine rates and economic status. Completion rates for the seven-vaccine series are substantially lower for children living below the poverty line than children living above the line. "Is there a difference in beliefs, or a difference in misinformation between these groups? Or is it more likely that there's a difference in structural circumstance—the ability to deal with the logistic and financial complexity of completing the entire series in a timely way?" Hornik asked. A study that Hornik and his associates did in poor areas of Philadelphia showed that individual beliefs about benefits and risks of vaccination were not important correlates of vaccination levels. What mattered was the structural circumstances people faced: long wait times at overburdened clinics, missed opportunities for vaccination, older siblings who needed the car, the cost of taking time off work, and so on. "Our advice to the city of Philadelphia was to forget about a communication campaign and fix the clinical system," said Hornik.

He recommended starting with the behavior of concern and considering misinformation as one among other competing explanations of that behavior. If it does turn out to be an important explanation, remediation can be considered. "Scientific misinformation may matter, but it does not always matter. We need to show that it does."

## THE INFLUENCE OF SCIENCE, HEALTH, AND CULTURAL LITERACY

Adding to the complexity of science communication is the problem of science, health, and cultural literacy, said Susan Scrimshaw, co-chair of the board of directors for the Nevin Scrimshaw International Nutrition Foundation and former president of the Sage Colleges. According to the National Center for Education Statistics, about one in seven people in the United States cannot read, and another 21 percent read below a fifth-grade level, she observed. The challenge is to empathize, be supportive, and form communities with people whose literacy is limited.

Low literacy makes communication about health and infectious diseases particularly difficult, Scrimshaw observed. Even most people with college educations are not going to understand such terms as diurnal, herd immunity, genetic load, endemic, or transmission. Further confusion can result when evidence leads to changes in findings and recommendations, as has happened recently with prostate cancer screening and with blood pressure treatment recommendations.

As an example of translation difficulties, Scrimshaw cited work she and her colleagues have done in Spanish-speaking populations where they used the term "riesgo," which is the literal translation of "risk." But communications research with the intended audiences showed that the word "peligro," or danger, resonated much more with women in that population. "We never would have thought of that from the psychology or public health perspectives. The women taught us how to communicate with them."

Language, culture, and education create variation in what people hear. Ethnicity and culture, language, gender, age, low literacy or low Internet literacy, economic status, blindness, and deafness all heighten the risk for miscommunication, said Scrimshaw. The risks of Zika infection are an example. People need specific information about the risks of infection and of a baby being affected. But the complexities of the behavior of mosquitoes make the task of conveying information difficult. Furthermore, when sexual transmissions proved to be occurring, discussion of condoms was a nonstarter in some political climates. Public health officials might assume that what they recommend is possible. But, for example, it may not be possible to avoid getting bitten by mosquitoes following a hurricane, when the power is out and buildings are damaged, which is what occurred in 2017 in the U.S. Gulf Coast and the Caribbean.

In general, people get information about health and infectious diseases in different ways than in the past. They have shifted away from newspapers and television to Internet sites and social media contacts. Judging credibility, developing trust, and understanding sources are all different for these media than for traditional media. Furthermore, culture, education, language, and other factors influence these processes, even as the ways people get information continue to change.

Scrimshaw drew several conclusions from her observations. First, communicators need to understand the cultural and linguistic parameters for the groups they are trying to reach. "One message does not fit all," she said. Iterative consultative and action relationships with communities help build trust. Telling stories that illustrate data and desired behaviors can be powerful influences on beliefs and behaviors.

Communicators also need to understand the feasibility of the behaviors they are seeking in terms of access, economic constraints, priorities, and other factors, said Scrimshaw. They then need to develop strategies to enable the desired behaviors. Social mobilization and community engagement can foster communication, as can integration of the social, communication, and health sciences. Long-term funding for capacity building and preparedness require political will and understanding. Finally, the role of the National Academies of Sciences, Engineering, and Medicine in building capacity and serving as science "advisers to the nation" is critical, she said.

# 4

# Creating a Collaborative Community

---

> **Important Points Made by the Presenters**
> - Communicating science effectively requires working across boundaries within and among research teams, institutions, and audiences. (Weingart)
> - Effective communication systems often involve boundary organizations that straddle and link political and scientific institutions. (Guston)
> - Collaborative development of a logic model can forge agreement on communication strategies and messages. (Tierney)

Effective science communication requires collaborations among diverse groups, said Frank Sesno, director of The George Washington University School of Media and Public Affairs, who moderated the first day of the colloquium. Subject-matter experts, social scientists, experts in eliciting and addressing audience needs, and practitioners who understand the interfaces of science, society, and storytelling all have roles to play. Collaborations can attract new researchers, institutions, and funders to the study of science communication while fostering sustainable institutional commitments to science communication research. Collaboration is a key ingredient, Sesno said, in providing "the scientific basis for effective science communication."

## MANAGING CONFLICTS IN SCIENCE COMMUNICATION

Laurie Weingart, interim provost and Richard M. and Margaret S. Cyert Professor of Organizational Behavior and Theory at the Tepper School of Business at Carnegie Mellon University, built on this theme by observing that communicating science effectively requires working across boundaries, whether within or among research teams, institutions, or audiences. Weingart has investigated interdisciplinary design teams, product development teams, project teams, research teams, and intensive care unit teams, and she drew on this work in addressing two broad questions:

- How does interdisciplinarity influence our ability to successfully work together?
- What can we do to bridge our differences and most effectively communicate with one another?

Interdisciplinary collaborations are challenging, she noted. People have different functional areas, expertise, past experiences, cultures, and affiliations, which the research literature refers to as different "thought worlds." Because people perceive and interpret situations differently, they arrive at different solutions. These diverse approaches can lead to better solutions to problems, Weingart observed, but they can also result in difficulties understanding one another, communicating with one another, and reaching agreements. If not managed appropriately, these difficulties can lead to ineffective collaboration and conflict.

As an example of a perceptual gap, Weingart cited the nurses and physicians in an intensive care unit. Nurses have the goal of symptom management, such as trying to help a patient experience less pain. Physicians are focused on treatment of the disease, with a goal of reducing reliance on a respirator. These two goals result in different definitions of improvement. "Is this about comfort? Is it about reliance on a respirator? This will lead to disagreements. It will lead to mixed messages to the family about what should be done and how this person should be treated, and it can affect the quality of coordination and decision making," Weingart noted.

Perceptual gaps can be bridged through cognitive integration, Weingart said, which relies on developing an understanding of another person's perspective. This can involve understanding a person's use of language, ways of reasoning, preferences and priorities, constraints, goals, and objectives. "The idea is that I don't have to know what you know, but I have to understand how you come at a problem so we can work together."

Perceptual gaps can also be bridged through affective integration, said Weingart. Trust can be gained through experience or reputation so that a person is willing to rely on or be vulnerable to someone else. Respect can be based on one's expertise, role, or status. Being liked can occur through on-the-job interaction, sharing friends, or socializing. These three factors do not always covary, Weingart pointed out. A person could trust someone else but not respect that person, or vice versa. A person could trust or respect someone else without liking that person. Liking may seem less necessary, Weingart added, but "it's difficult to get your work done if you don't want to be in the same room with other people."

When interdisciplinary collaborations have difficulties, they can generate conflict, but is conflict good or bad for teams? Research shows that conflict will surface disagreements and can spur innovation. At the same time, conflict can disrupt productivity and the ability to listen to one another. The type of conflict is one factor in determining outcomes. Is it conflict about the task itself, or is it about the relationships among group members? Another factor is whether conflict is managed collaboratively or competitively. "There's isn't an easy answer. We used to think that task conflict was good, and relationships conflict was bad," Weingart said. But the literature has demonstrated more nuanced effects, depending on how conflict is expressed and managed. "Conflict management matters, and we need to understand this dynamic better."

Weingart identified two dimensions in the expression of conflict. One is the directness of the conflict: how explicit are the opposing positions that are being conveyed? The other is the intensity of the conflict: what is the degree of force with which opposition is being conveyed? These two factors can vary independently, yielding four general categories of conflict expression. When conflict is expressed with high directness and high intensity, arguments ensue, which may cause people to quit listening to each other. Conflicts expressed with high intensity and low directness often take the form of undermining others, whether through complaints, backstabbing, or teasing. With low directness and low intensity, people may disguise or avoid conflict, which can lead to uncertainty and stress. With high directness and low intensity, they are more likely to discuss the issues. "With this approach, you can get a sense of debate being a dominant communication style for conflict," said Weingart. "You're avoiding conflict spirals, increasing information, and decreasing emotion. . . . It also increases the positive emotion in terms of energy and excitement." For these reasons, conflicts expressed with low intensity and high directness are most likely to result in high-quality agreements.

## BRIDGING BOUNDARIES IN SCIENCE COMMUNICATION

At Arizona State University, David Guston has been director of the large National Science Foundation–funded Center for Nanotechnology in Society and is currently director of the School for the Future of Innovation in Society. Both of these organizations have had a great deal of internal diversity—of the 34 people with Ph.D.s at the School for the Future of Innovation in Society, for example, 28 were from different disciplines. From these experiences, Guston has concluded that managing interdisciplinarity is about building relationships. In this way, work at the centers has been able to take advantage of not just individual faculty members' skills and talents but also their interactions among themselves and with others.

Interdisciplinary science communication can take many forms, Guston observed. Books, video, audio, or games can all facilitate two-way interactions. Informal and out-of-school education can provide a venue for science communication. The fine and performing arts are another way of conveying scientific information, as is policy programming and community engagement. However, all of these approaches require policies, academic leaders, and colleagues who value these activities.

Guston cited a study of sustainability science (Cash et al., 2003) that identified three key issues for science communication: saliency, credibility, and legitimacy. Saliency relates to relevance. "Is the research framing done in a way that is relevant to the decision makers? Does it have their questions in mind?" Credibility relates to whether the information can be trusted. "Does it come from an esteemed and respected [source]?" Legitimacy relates to whether the communicated information represents something that is politically or sociologically legitimate to its recipients. "Can they understand the array of interests behind it? Can they see that some consideration of political nuance and sensitivity was put into the structuring of the research?"

Effective communication systems often involve what Guston termed boundary organizations (Guston, 2000). Boundary organizations straddle the political institutions that make public policy decisions and the scientific institutions that gather and disseminate empirical evidence. These boundary organizations have opportunities and incentives to use boundary objects, such as patents, that are valued on either side of the boundary. Boundary organizations also involve participation of actors on both sides of the boundary as well as mediating professionals. This reflects the fact that engagement is a bidirectional process in which people from both sides engage with each other in a protected space, with assistance from individuals who are skilled at communication and collaboration. Finally, boundary organizations have distinct lines of accountability to each side. Some of their work is recognized as important to scientific organizations, and some of their work is recognized as important by political organizations. "Boundary organizations give both producers and consumers of research an opportunity to construct the boundary, to construct how their domain differs or is similar to the other enterprise, in ways that are favorable to their own perspective. Therefore, these things may persist over time," Guston concluded.

## COLLABORATIVE SCIENCE COMMUNICATION RESEARCH

In thinking about barriers to interdisciplinary collaboration, Kathleen Tierney, professor of sociology at the University of Colorado Boulder, introduced the concept of hierarchies of credibility, which was developed by sociologist Howard Becker in the 1960s. This concept refers to the taken-for-granted assumption that those with high status have the right to define how things really are, while those of lower rank have incomplete information or are misinformed. These hierarchies of credibility in science pose tough communication challenges, said Tierney. Scientific disciplines have different levels of prestige, with disciplines like mathematics and physics at the top of the hierarchy and other disciplines trailing behind. Even within disciplines, some subdisciplines have more prestige and have attracted more attention and respect than others. Applied research and basic research affect perceived levels of prestige, as does the ranking of an academic institution. "And of course all that is overlaid across the pervasive significance of gender and of race when team members are communicating with one another," she said.

As an example of collaborative research on science communication and associated evaluation efforts, Tierney briefly described a collaboration between the U.S. Geological Survey (USGS) program on Science Application for Risk Reduction (SAFRR) and the University of Colorado Natural Hazards Center. The SAFRR program has created scenarios of future disasters with accompanying materials, such as videos and animations, to try to communicate to stakeholders and the general public what to expect in the event of a disaster. One of these was called the Great Southern California ShakeOut, which was based on a scenario for a major earthquake on the southern part of the San Andreas Fault. ARkStorm dealt with a massive flood in the northern California Delta caused by an atmospheric river, while Haywired centered on a scenario for an earthquake on the Hayward Fault in the San Francisco Bay Area. Development of these scenarios, which was an "extremely labor-intensive" process, according to Tierney, involved people from a variety of disciplines, including economics, public health, geology, and engineering.

After the ARkStorm scenario, USGS brought in social scientists from the Natural Hazards Center to evaluate a scenario involving a tsunami striking coastal California. Tierney and her colleagues at the center first developed a logic model that posed such questions to the scientists as, "What are the components of the communication strategies that you're going to be using? What are the outcomes that you expect from these strategies?" According to Tierney, "it was incredibly hard to get agreement among the scientists on why exactly they were doing this form of communication." The researchers then conducted online surveys with stakeholders involved with the development of the scenario, including personnel at the ports of Los Angeles, Long Beach, and San Francisco, representatives of the California Department of Transportation, and emergency managers. The survey asked

- Was the scenario development process credible?
- Did stakeholders feel engaged in the process?
- Did their understanding of tsunamis and potential impacts increase?
- Did they seek out additional information after meetings?
- Did they pass on what they had learned to others?
- How useful and applicable was the information provided in the scenario?

Another aspect of the project involved the production of a public service video. In a pretest/posttest design, the video, which was a collaboration between USGS and the Art Center College of Design in Pasadena, was evaluated through focus groups that answered questions before seeing the film, after seeing the film, and 4 weeks later on three topics: tsunami characteristics, tsunami warning signs, and self-protective actions. Questions were both close ended and open ended, with opportunities for feedback.

Many viewers, including young people, thought the video was most appropriate for younger viewers. Some people who had relatives in areas that had been affected by tsunamis did not think the humor in the video was appropriate. There were no differences between people who saw the video once and people who saw it three times. But there were differences between people of different ethnic groups and educational levels.

# 5

# Incentives in Science Communication

---

### Important Points Made by the Presenters

- The incentive structures in higher education are changing in ways that support science communication, though much more needs to be done. (Hoffman)
- The establishment of identity and socialization, the development of agency and self-efficacy, and recognition systems all provide opportunities to integrate science communication into faculty careers. (O'Meara)
- Revision of tenure guidelines could move toward integrated systems that combine elements of public outreach, traditional tenure criteria, and a candidate's overall impact on a field. (Scheufele)
- Dialogue among the stakeholders in a community can establish a common vision, identify barriers for achieving that vision, and brainstorm pathways for overcoming those challenges. (Taylor)
- This dialogue needs to include people in industry and in national laboratories. (Donahue)
- Interdisciplinary collaborations can strengthen the communication capabilities of all the members of a collaboration. (Skop)

Incentive structures in higher education have a major impact on how scientists communicate critical advances in their fields. As Andrew Hoffman, Holcim (US) Professor of Sustainable Enterprise at the University of Michigan, said, the rewards system is a statement "about who we are, what we do, why we do it, and what we value. If we don't change the rewards [system], nothing is going to change."

Many academic scientists believe that it is important for the public to know more about science but think that they will not benefit personally by engaging in science communication, noted Hoffman. According to a poll of University of Michigan faculty conducted in 2015, 56 percent thought that public engagement is not valued by tenure committees, 41 percent thought it was time consuming and distracting, and 34 percent believed that public engagement is dangerous because sources are often misquoted. However, the poll also showed that younger scholars want more engagement. "They want to change the world around them," Hoffman said.

He distinguished four combinations of motivations and incentives as a two-by-two matrix:

*Extrinsic Motivation and Formal Incentives* encompass tenure and promotion, annual reviews, journal publishing, and training and include much of what is considered the rewards system in academia.

*Extrinsic Motivation and Informal Incentives* include gaining greater visibility within a field or among the public, which can improve a person's status within the university. However, such a person can also be mocked or dismissed as a popularizer or media hound who is simply trying to attract attention.

*Intrinsic Motivation and Formal Incentives* can lead a scientist to develop new research strategies or seek new audiences as a way to explore new and novel approaches and increase stature and impact both inside and outside the academy.

*Intrinsic Motivation and Informal Incentives* can provide a person with a sense of meaning and purpose. Scientists may do something "because they believe that's what they're supposed to do, it's their purpose, it's their mission."

Hoffman pointed to several signs that the rewards system is changing for the better. High-level administrators are beginning to pay more attention to the issue, though "I'd love to see more presidents, more deans, more provosts, more funders, the people who really have their finger on the institutions" engaged in such discussions, Hoffman said.

In addition, more institutions are getting involved in the issue. For example, a recent report from the American Sociological Association explored how to evaluate public communication in tenure and promotion (ASA, 2016). The Ross School of Business at the University of Michigan added a fourth category—practice—to the three traditionally included in annual reviews—research, teaching, and service. The Mayo Clinic now includes social media scholarship activities in tenure and promotion decisions.

Finally, journals are beginning to change. One initiative, for example, is encouraging business management journals to focus on more issues of empirical relevance. "It's much easier to engage in public discourse if you're doing research that the public actually cares about," said Hoffman.

## FACTORS THAT INFLUENCE FACULTY BEHAVIOR

A number of factors predict faculty behavior, said KerryAnn O'Meara, professor of higher education at the University of Maryland, College Park. She focused on three (at the risk, she admitted, of oversimplifying): identity and socialization, agency and self-efficacy, and recognition. With all three, different groups could integrate science communication in "sticky" ways into faculty careers.

Identity and socialization constitute part of the process by which people become academics. In graduate school, students acquire knowledge, skills, and orientations within their disciplines and institutions. Thus, if graduate students do not get opportunities to develop the knowledge and skills of science communication and do not see this work as being valued by their academic mentors and peers, they will not develop an orientation toward it.

As an example of one way to socialize future faculty toward this type of work in graduate school, O'Meara cited the University of Maryland's Language Science Center. As part of a National Science Foundation (NSF) Research Traineeship program, the center aims to train graduate students to communicate about their research through policy internships, opportunities to meet with policy makers, writing workshops, interactions with professionals in the field, and outreach to high school linguistic classes. Students in this program practice explaining their work and receive feedback to improve their approach and style. They also receive strong role modeling and signals from their faculty that "science communication is part and parcel of being an excellent scholar," said O'Meara.

By agency and self-efficacy, O'Meara said that she was talking about strategic perspectives and actions to accomplish the goal of communicating with public audiences. Despite many innovations in graduate education, most faculty today are trained to study specific phenomena and then communicate that knowledge through presentations and journal articles. With support from federal agencies, foundations, and other organizations, groups are working with faculty members to broaden their targets through such means as blogs, op-eds, national and local interviews, and social media posts. For example, the OpEd Project is a leadership program with the mission of trying to change who narrates history and interprets findings. "Who's invited to the research panel to write an op-ed? Who gets to be on NPR to weigh in? They've been very successful at bringing in new voices to public conversations," O'Meara said.

Finally, recognition includes promotion and tenure and the other ways that academics are rewarded. "I like to think of reward systems more as *regard* systems, because they're about much more than promotion and tenure," said O'Meara. In the process, biases and existing reward systems can privilege some voices over others, limiting the number of people at the table for important conversations. But criteria for evaluating scholarly activity have begun to change. For example, many institutions have begun to create awards and track various kinds of high-impact engagement. The institutions that have been most successful have been those that have integrated activities into a holistic portfolio, avoiding the faculty members' fear that public engagement will be an add-on responsibility that everyone must do. This approach calls for clear criteria and the use of language within that field to identify high-quality work, O'Meara said. "It's a matter of changing the view that we have of the work to move it into a form of scholarship that can be acknowledged for what it is."

## AMBASSADORS FOR SCIENCE AND ENGINEERING

"As an engineer, my focus is on taking scientific understanding and using it to create societal change," said Emmanuel Taylor, senior electricity consultant at Energetics, Incorporated. "It is impossible to do that unless technical professionals collaborate and communicate with broader audiences."

As a graduate student, Taylor was involved in the Science and Engineering Ambassadors program, an initiative of the National Academy of Sciences and the National Academy of Engineering to better equip scientists and engineers to be able to communicate and then to create dialogues between scientists and engineers and the broader communities in which they live. Similarly, as a technical consultant for Energetics, Incorporated, he works with public- and private-sector organizations to create technology road maps and strategic plans. These often require direct dialogue among stakeholders in a community to establish a common vision, identify barriers for achieving that vision, and brainstorm pathways for overcoming those challenges. "Both at the community level and the industry level, we've seen the benefits where communities are able to utilize scientific understanding to make complex decisions, and industries are able to move forward with clearer vision and purpose, in order to create benefits that have a measurable impact beyond publications," he said.

Scientists and engineers benefit in many ways from engaging in these dialogues, said Taylor. They are exposed to opportunities that can advance their careers. They can affect policy and regulation by making available and interpreting data. Their work can also have a greater impact on society. "Often researchers will spend years developing new technologies and then attempt to commercialize it, only to find that industry is working with a different set of assumptions," he said. "That can be overcome by direct dialogue and communication."

Taking early-stage research to commercialization in a complex socio-economic environment requires the coordination of scientists, engineers, and other professionals from universities, national laboratories, private industry, and other organizations, Taylor said. He has been helping Energetics, Incorporated, structure those dialogues and create forums, summits, and workshops to gather diverse opinions. "Increased understanding and knowledge, and advanced technology, can have a measurable impact on the quality of life for communities and can change advanced industries," he said.

As an example of how the Science and Engineering Ambassadors program has been used as a vehicle to engage the public, Neil Donahue, Lord Professor of Chemistry at Carnegie Mellon University, described his experiences publishing several papers on climate change. They were on the formation of fine particles in the atmosphere, which is a major source of uncertainty in climate forcing. This work was picked up by *The Wall Street Journal* and used to suggest that computer models have overstated the risk of a warming climate. But this was not at all what the papers showed, said Donahue. Rather, they showed that fine particles could slightly cool the atmosphere to an extent well within the uncertainty range estimated by the Intergovernmental Panel on Climate Change. "We weren't disagreeing with anything," said Donahue.

Since then, Donahue and his colleagues have been working with Baruch Fischhoff and Illah Nourbakhsh at Carnegie Mellon University to engage communities in Pittsburgh around the questions about air pollution. The Science and Engineering Ambassadors program has been particularly valuable in this effort, he said, because it has engaged people in industry and in national laboratories "to get the dialogue going."

## REVISING TENURE GUIDELINES

Tenure guidelines differ among institutions, but they typically require demonstrated excellence in research and in either teaching or service, which at land grant universities has often included extension activities. But recent years have seen pushback against these criteria even at major research universities, said professor of science communication Dietram Scheufele, who chaired a committee that looked at a revision of tenure guidelines at the University of Wisconsin. That committee recommended what he called an "integrated case" that combines elements of public outreach, traditional tenure criteria, and a candidate's overall impact on the field. In particular, working with communities and other stakeholders would become a larger part of a tenure package.

However, questions of how to quantify the quality of such activities have long plagued tenure committees, Scheufele added. For example, a major problem is distinguishing between research and the impact of that research. Can the same activities count toward both categories at the same time? What are the criteria to be used for excellence? Are standards being watered down?

Scheufele said that such challenges can and should be seen as opportunities. Focusing on outcomes provides an opportunity to clearly define outreach, engagement, communication, and public scholarship. In the past, these terms have not been well defined, which complicates the writing of tenure guidelines. "We need to do a lot more thinking in that respect and we have to be very clear."

Greater clarity also will yield greater transparency, he said. The criteria for promotion can be established when people are hired and then people can be evaluated against those standards. For example, the standards at the University of Illinois at Urbana-Champaign, call for independent, verifiable, and specific evidence of excellence and impact in the area of outreach.

The University of Wisconsin now has a Ph.D. minor in science communication for bench scientists that shows up on their transcript. Such training prepares bench scientists for careers that involve outreach, which has been particularly welcomed by younger scientists. These young researchers have a hunger to make a difference, Scheufele said. "It's a generational shift."

## MELDING ART WITH RESEARCH

Ahna Skop, associate professor in the Department of Genetics at the University of Wisconsin–Madison, studies cell division, which she described as a "beautiful" process. With an affiliate faculty position in both life sciences communication and the Arts Institute of Wisconsin, she has been doing outreach and engagement through scientific art and trains artists in her laboratory. Art provides creative and innovative ways of visualizing and tackling problems, she said. For example, partly because she is a visual learner, she was transfixed by the process of cell division as a young researcher. She also became fascinated by what she saw underneath the microscopes of other people. As a graduate student, she started the International *C. elegans* art show, which has been going on for 20 years in the *C. elegans* community. She also noticed that the walls of the laboratory at the University of Wisconsin had no art. "So I went to the dean's office, above my chair, and I said, 'I have an idea. I want to put art on the walls because I want students to come in and see what we're doing.'" With the $15,000 she got from the dean, she has filled the department with art. "These are images that engage the public and bring them in," she said. In turn, this led to the Cool Science Image Competition at the University of Wisconsin, in which scientific images are displayed as art and are a way of interesting schoolchildren and other members of the public in science. "As someone who takes funding that is from taxpayer money, it's an obligation to give back to society," Skop said.

She pointed to the broader impacts criterion introduced into the grants process at the NSF as a powerful incentive for these kinds of programs. One way for scientists to have broader impact, she said, is to integrate the arts into their work. "If you have a passion or hobby, you can combine it with what you do," she said. "You can meld that with your science, and there are lots of funding mechanisms to support this."

---

### The Science Behind the News: Human Gene Editing

New techniques have made it possible to alter the DNA sequence in a cell with single-letter precision, explained Matthew Porteus, associate professor of pediatrics at Stanford University. Furthermore, these new techniques have not only revolutionized but democratized the field of genome editing, because they are relatively simple and inexpensive to use. "It's not quite do-it-yourself in the garage, despite the advertisements, but essentially any lab with reasonable molecular biology experience can do this," said Porteus.

Porteus and his colleagues have been using the procedure in an attempt to cure sickle cell disease, which is caused by a single change in the nucleotide sequence of the gene that encodes a protein called beta-globin. They have edited the gene sequence in the stem cells that produce red blood cells over a person's lifetime, correcting the misspelling that causes the disease. The technique has been a success in the laboratory, and at the time of the colloquium the researchers were working

*continued*

with the U.S. Food and Drug Administration to develop a clinical protocol to begin treating patients with the disease.

Human gene editing raises difficult ethical issues, Porteus said. Initially, the technology will be used with a very small number of people in the countries with the strongest economies. But the disease affects people around the world and is especially concentrated in Africa. "How do we get sophisticated technologies to parts of the world that don't have the same GDPs [gross domestic products] as we do?" he asked.

Porteus also distinguished between using genome editing to cure or treat a serious disease and using it to enhance human traits. Though the line between treatment and enhancement is not always sharp, an ethical distinction exists between treating sickle cell disease and creating dystopian soldiers or babies with particular appearances.

He drew the distinction between somatic cell editing, in which case the genetic change is limited to a single person, and germline cell editing, in which case someone could pass a genetic change on to offspring. "They're not the same thing," Porteus argued.

Finally, he pointed to several "perverse incentives" that surround the communication of scientific information in general. All stakeholders, except the public, are rewarded by sensationalizing science. Researchers are more likely to get promoted and funded if their work is featured in *The New York Times*. Journals are more likely to publish their work if it is sensationalized because they are interested in increasing their revenue or status. Companies are more likely to be able to raise funds or have their stock price increase. Journalists are also more likely to be read or viewed when they sensationalize or overdramatize the discoveries of science. "Where is the check in this process? I don't have an answer," he said.

The story of human gene editing is difficult to tell, acknowledged Cornelia Dean, former science editor of *The New York Times*. The topic is technically complex. The concepts can be difficult for journalists and readers who are not specialists to understand. Furthermore, the story raises fundamental questions about what it means to be human, which can be a problem when many people believe that science and technology are moving too quickly and that insufficient attention is being paid to moral and philosophical issues.

In December 2015 the Organizing Committee for the International Summit on Human Gene Editing, which was an international effort involving the National Academy of Sciences, the National Academy of Medicine, The Royal Society, and the Chinese Academy of Sciences, released a statement that it would be irresponsible to proceed with germline gene editing unless a broad societal consensus is achieved about the appropriateness of this work and its proposed applications. The committee said that a wide variety of people should be involved in these discussions, including biomedical scientists, social scientists, ethicists, health care providers, patients and families, people with disabilities, policy makers, regulators, research funders, faith leaders, industry, members of the general public, and public citizen advocates. "How we could manage to pull those kinds of conversations together in our country today, I have no idea," said Dean. "But that's what is needed," and not only on human gene editing but on other difficult subjects, such as geoengineering, autonomous robots, and artificial intelligence. "The research community needs to reach out to the wider society and try to get this kind of conversation going."

Dietram Scheufele, John E. Ross Professor in Science Communication in the Department of Life Sciences Communication at the University of Wisconsin–Madison, pointed to another consideration. In general, the wealthier a nation is, the less religious it tends to be, but the United States is a prominent outlier. More than half the people in the United States say that religion plays a very important role in their lives, compared with less than 20 percent in Britain, China, France, Japan, and Russia. "That's not good or bad, that's just a reality," said Scheufele. "Political and societal discourse about technologies will take place in this environment."

A religious lens often colors how people in the United States view new technologies. For example, Scheufele pointed out, survey respondents are more likely to think that synthetic biology conflicts with moral or religious views and blurs the line between humans and God than is the case with nanotechnology (Akin et al., 2017). When asked whether the scientific community is guiding the development of human gene editing responsibly, respondents had more abstract concerns—that the technology can be "easily used for the wrong purposes" or "messes with nature"—than they did about more concrete concerns—the use of the technology to remove stigmas around birth defects and disease (Scheufele et al., 2017). The more religious the respondents are, the more likely they are to think the scientific community is not capable of handling new technologies responsibly by itself. Greater knowledge about an issue produces the opposite effect: the more people know, the more likely they are to believe that science will be able to roll out new technologies in a responsible fashion.

Scheufele concluded that "regardless of what we might think or what our initial feelings are toward a technology . . . we all want to have these broader conversations." A complication is that different groups of people have varying levels of trust in other groups. Thus, more religious survey respondents are less likely to trust scientists, more likely to trust parents, and much more likely to trust religious groups than are less religious respondents.

# 6

# Communicating with Policy Makers

> **Important Points Made by the Presenters**
> 
> - More information on how to communicate and engage with policy makers is a common request of science communicators. (Cloyd)
> - Reviews of existing research, collections of best practices, interviews with scientists and policy makers, and investigations of macroscale influences on policy makers' use of scientific information could yield much more informed guidance for science communicators. (Suhay)
> - Updated and expanded science policy training programs could help scientists communicate effectively and ethically with policy makers and lead to deeper discussions from both sides of how science policy and science-informed policy is made. (Nash)

In workshops that the American Association for the Advancement of Science (AAAS) has held for scientists since 2008 on communicating research findings, a frequent request has been for more information on how to communicate and engage with policy makers. In 2016 the AAAS began work on a new module for the workshops that would meet this request, said Emily Cloyd, project director for public engagement at the AAAS. The organization began by bringing in Elizabeth Suhay, assistant professor of government in the School of Public Affairs at American University, and Erin Nash, a doctoral candidate at Durham University in the United Kingdom, to produce a literature review that would inform the new module. The work on the literature review revealed that "not much is known about how policy makers, as a specific subpopulation, form their beliefs about science or science information," said Nash. In the *Communicating Science Effectively* report, for example, just 4 of 100 pages are devoted to the topic of communication with policy makers, and the report points to the paucity of research on this subject. Furthermore, policy makers and the general public have some important differences in how they engage with scientific communication. First, policy makers are more exposed than the average person to multiple channels of information, including actors with deep political motivations. Second, elected officials have greater access to sources of information from specialists, such as the testimony of experts and information analyses provided by the Congressional Research Service. Third, representatives and their staff routinely consider political strategy when forming their policy preferences. Fourth, the decisions of policy makers are one of the most powerful influences on the general public's beliefs about scientific issues and their attitudes toward science-informed policy. "Attention to policy makers as a group is an especially important gap to fill," she said.

However, the ability to fill this gap has been limited, said Cloyd, because few sources of support have been available for this kind of work. Before the third Arthur M. Sackler Colloquium on Science Communication, the National Academy of Sciences, with support from the Rita Allen Foundation, announced a competition for proposals from research-practitioner partnerships to investigate the priority topics identified in *Communicating Science Effectively*. The project led by Suhay was one of two that were selected from the submitted proposals. (The other is described in the section "Countering Vaccine Hesitancy" in Chapter 3.)

Recommended practices from the project, which is just getting under way, will be partly based on several stages of empirical research. Building on the literature review, the team will summarize research on the communication of science more broadly where it might apply to policy makers. Two key types of literature are being reviewed: the academic literature, and literature from policy, science, and science communication–based organizations, such as guidance material that has been developed to provide scientists and communicators with advice on how best to communicate science to policy makers.

The second component of the project will use a survey to better understand the practices of those who communicate science to policy makers. A representative sample of AAAS members, as well as, more narrowly, people within policy and government relations teams of the 250 scientific organizations that are affiliated with the AAAS, will be surveyed to answer the following questions: Who is most likely to engage with policy makers? Who is communicating with whom? What information is communicated? How is it communicated, and with what goals in mind? What ethical concerns are involved? What defines success or failure in these interactions?

The third component is an influence-mapping study that will aim to provide a better understanding of how macroscale structures and environments, rather than just microlevel practices, affect policy makers' perceptions, understanding, and use of scientific information. These maps will describe the pipelines through which information, knowledge, and sometimes misinformation flow to pinpoint key points in the process.

In the final component of the project, 20 to 30 members of the U.S. Congress and congressional staff will participate in semistructured qualitative interviews that explore their perspectives on science communication. The team will draw from their networks to target and contact staff members and lawmakers. Interviewees will be selected based on several criteria, including partisan balance, level of scientific engagement in their congressional committee, and the degree to which scientific evidence is relevant to the legislation they deal with on a regular basis. The team hopes to enlist a mixture of participants, including both advocates and skeptics of science.

In these interviews, participants will discuss their viewpoints on contrasting case studies: those with contested or politicized scientific claims, and those with lower levels of contestation. Members will be asked whom they rely on for information and their rationales for trust. Time permitting, the team hopes to interview staffers of the Government Accountability Office, the Congressional Research Service, and the Congressional Budget Office, all of whom regularly provide science-based information to members of the U.S. Congress and their staff and who are under professional obligations to present nonpartisan and objective information and evidence.

The project aims to share its findings through scholarly articles; a report to the National Academies of Sciences, Engineering, and Medicine; and blogs. The AAAS will use the findings to update and expand its science policy training programs for scientists, engineers, and students; to expand the resources available in their fellowship programs, including the Leshner Leadership Institute for Public Engagement with Science and Technology; and to share materials with its 250-plus affiliate societies. In addition, the project has been designed to be mutually beneficial to policy makers and scientists, Nash said. Its recommendations can guide scientists in effective communication with policy makers while policy makers can better understand how their opinions are formed and spot potential blind spots. Results could lead the way to deeper discussions from both sides on how scientific policy is made.

## CONGRATULATIONS AND CAUTIONS

All three of the discussants of the project praised its objectives and design. The project could "create communication strategies based on facts instead of gut instincts," said James Cohen, director of communications and public outreach for The Kavli Foundation. Focusing on the U.S. Congress as a specific audience will help clarify approaches and messages. The study will also lay the groundwork for subsequent studies that could investigate more specific questions, such as how the scientific community and policy makers in specific districts or states interact.

David Herring, program manager of the Communication, Education, and Engagement Division at the National Oceanic and Atmospheric Administration's Climate Program Office, observed that members of the U.S. Congress value science but often on a selective basis that aligns with their motivations and agendas. Thus, a major research question is how their agendas color their ability to receive, hear, and understand science. In addition, differences in culture, values, faith, and perceptions of risk may influence policy makers' interactions with scientific information. Policy makers' unique characteristics call for research that recognizes their unique position in relation to science and the scientific community, Herring said.

The project will not be without challenges, noted Fay Cook, assistant director for the Directorate for Social, Behavioral and Economic Sciences at the National Science Foundation. Response rates are always problematic with survey research and can be especially difficult when surveying elites. The year-long timeframe planned to conduct the three sequential studies will be equally challenging. Cook recommended ensuring that each study focuses on the same questions and that enough time is left after each study for analysis and subsequent planning. Finally, the project will have to resist the broad, sweeping conclusions that often result from this type of work in favor of specific, evidence-based recommendations, she said. Ideally the three sequential studies will give rise to a contingency theory of science communication that can be further honed and tested: "Under what conditions, with what kind of issues, and for whom can certain kinds of science communication make a difference?"

# 7

# Threats to Science's Reputation

---

> **Important Points Made by the Presenters**
> - The self-correcting processes of science may be fueling a new narrative that science is broken, even though scientists are the ones uncovering problems and moving to solve them. (Jamieson)
> - Scientists and engineers are seen as highly competent, but many members of the public have no opinion about whether they are trustworthy. (Fiske)
> - Deviations from best practices that increase the likelihood of error and vulnerability to misconduct can reduce trust in the reliability of the scientific process. (Finneran)
> - Standards developed by the research community to support norms of sharing data, facilitating review and replication, and reducing bias can enhance confidence in science. (McNutt)

Research misconduct, irreproducible results, retractions, conflicts of interest, and other issues surrounding the scientific process can erode public trust in science and scientists. To what extent do these violations of research integrity threaten the reputation of science? How can they be distinguished from the self-correcting nature of science, which tries to replicate and falsify existing findings in the pursuit of new knowledge?

Furthermore, what are the best ways to describe and explain these issues so that people better understand the context and norms of science?

## THE "SCIENCE IS BROKEN" NARRATIVE

The dominant narrative in the media's coverage of science is the quest for discovery, observed Kathleen Hall Jamieson, Elizabeth Ware Packard Professor of Communication at the University of Pennsylvania's Annenberg School for Communication. This narrative is not about problems, crises, or science being self-correcting. However, the self-correcting processes of science may be fueling a new narrative that science is broken, Jamieson observed.

Analysis of science news articles from *USA Today*, *The Wall Street Journal*, *The New York Times*, and *The Washington Post* from May 2016 to April 2017 revealed that the articles tend to contain words like *breakthrough*, *advance*, *path breaking*, and *paradigm shifting*. Most included a discovery or finding, credited the involved scientists or institutions, and noted the significance of the finding. Few mentioned past failures or false starts. The result, said Jamieson, is a mistaken impression that science is a linear process that goes from intuition to discovery without intervening difficulties.

A potential problem with the dominant news narrative is that not covering false starts or tentative conclusions that later change may give people the impression that difficulties are an aberration in the scientific process. *The Economist* of October 18, 2013, headlined such an article: "Trouble at the lab: Scientists like to think of science as self-correcting. To an alarming degree it is not." Readers of such an article can easily miss the fact that scientists found those problems, disclosed them, and were in the process of correcting them, said Jamieson.

In a study of the "science is broken" narrative, Jamieson and her colleagues conducted a search of news databases for the term "science" near any of the terms "crisis," "broken," "problem," "self-correction," "retraction," "replication," "peer review," "scandal," and "fraud/fake" from April 2012 to April 2017. They found 121 articles and opinion pieces, which, after duplicates and unrelated stories were eliminated, yielded 76 articles for analysis. Of these articles, 31 focused solely on a new scientific finding about a problem in science, 26 were authored by a scientist, 22 noted that science is self-correcting or made a comparable statement and recommended at least one solution, and 5 indicated that the problem is exaggerated or not real. The overall conclusion was that the articles identified scientists as the ones uncovering problems but largely failed to note that they are also the ones moving to solve them. In short, scientists are playing a role in perpetuating the "science is broken" narrative, Jamieson said.

## ENHANCING TRUST IN SCIENCE

The "science is broken" narrative hinges on the issue of trust, observed Susan Fiske, Eugene Higgins Professor of Psychology and Public Affairs at Princeton University, and trust is complicated to maintain. Trust is crucial and adaptive, but it is also relative and fragile. Trust improves joint outcomes, increases effectiveness and accuracy in judging risks, and helps society progress. Societies with higher baseline levels of trust "are more economically successful because you trust that a stranger is going to live up to the contract," said Fiske.

For scientists, establishing trust is crucial to communicating credibility, said Fiske. This credibility rests on both expertise, which people believe that scientists have, and trustworthiness, which depends on perceptions of a person's motivation to be truthful. These perceptions are more of a gut response, Fiske noted, as opposed to evaluating expertise.

To the extent that people see scientists as being on one side of an issue in a polarized climate, trustworthiness will be a consideration. In that case, people are going to ask, for example, Are you on my side or not? Are you a friend or a foe? "If you're on my side and you share my goals, then you're warm and trustworthy," Fiske said. People make this decision rapidly and do so in similar ways over different cultures and across time.

Expertise (or competence) and warmth are orthogonal in that they vary independently and together form a two-dimensional space that makes intuitive sense (Fiske and Dupree, 2014). For example, nurses, teachers, and doctors are seen as being high in competence and warmth, whereas laborers, customer service agents, and fast food workers are seen as being low in competence and warmth. Scientists and engineers are seen as highly competent but only moderately warm. This moderate score on the warmth scale, Fiske suggested, is because people do not know whether scientists are trustworthy or not. If scientists are viewed as elites, they could be seen as exploitative. "The downside risk for scientists, if we don't get this right, is that people resent elites and dehumanize elites as unfeeling machines," Fiske said. "That would be a bad place for us to end up."

For now, public confidence in science remains high and stable, according to polls. "But compared with whom?" Fiske asked. "We are better than lawyers and CEOs—and better than Congress—but worse than doctors, teachers, and nurses, whose motivation, people think, is to be in their professions to help other people. We're about the same as accountants and the military."

Fiske and her colleagues have taken a different approach to gauge the perceptual trustworthiness of scientists. They asked people what attributes spontaneously came to mind when they thought about scientists. The most frequently cited attributes were smart, intelligent, educated, and curious, followed by such terms as thoughtful, organized, rational, nerd, male, and disciplined (Nicholas and Fiske, in preparation). But almost none of the terms involved trustworthiness and warmth. "This suggests that we're coming out in the middle of the trustworthiness-warmth scale because it's absent," Fiske pointed out.

Perceptions of trust are more easily lost than perceptions of competence, Fiske emphasized. "Think about a close relationship. If one person betrays the other person, it's really hard to come back from that." Furthermore, when described as competent, people may be passing judgment on the trustworthiness of sciences, in the same way that describing someone as "nice" may cast innuendos on their competence.

"Our reputation is at stake," Fiske concluded. Remaining trustworthy requires consistency over time in safeguarding perceptions of reliability and trustworthiness. Paraphrasing Jamieson, Fiske noted that "when there's no narrative, you'd better find one, because otherwise somebody else might find it."

## ADDRESSING DETRIMENTAL PRACTICES IN SCIENCE

Kevin Finneran, editor-in-chief of *Issues in Science and Technology*, agreed that "trustworthiness is something we have to be responsible for ourselves. We have to look at all the things we can do within the research process and the operation of science to make our work as reliable and believable as possible."

In 1992 the National Academy of Sciences (NAS), the National Academy of Engineering, and the Institute of Medicine published the report *Ensuring the Integrity of the Scientific Research Process*, which focused on threats to the research process from fabrication, falsification, or plagiarism, collectively termed "scientific misconduct" by the report (NAS et al., 1992). A quarter century later, the National Academies of Sciences, Engineering, and Medicine published *Fostering Integrity in Research*, which revisited many of the issues covered by the previous report (NASEM, 2017c). *Fostering Integrity in Research* noted that the research enterprise has changed markedly in the quarter century since the earlier report. It is larger, more complex, more regulated, more dependent on information technology, more oriented toward commercialization, more international, and more relevant to policy decisions. All of these changes make maintaining the integrity of research more challenging, noted Finneran. Collaborators from other countries may have different standards of review and responsibility. Researchers from different disciplines may have different ways of listing authors or making data available. "There are a variety of things that we have to stay on top of and negotiate," he said.

Both reports noted that reliable data about the extent of scientific misconduct are difficult to gather. The number of misconduct cases identified by the National Science Foundation and the National Institutes of Health (NIH) has remained small. However, in a survey of research psychologists, between one-quarter and one-half admitted engaging in practices such as failing to report all of a study's conditions, selectively reporting only studies that "worked," or reporting an unexpected finding as having been predicted from the start (John et al., 2012). Another survey of researchers with funding from NIH found that 10 percent admitted practices such as inappropriately assigning authorship credit or withholding details of methodology or results in papers or proposals (Martinson et al., 2005). A meta-analysis of researcher surveys concluded that 2 percent admitted to falsifying or fabricating data at least once and 14 percent were aware of a colleague doing so (Fanelli, 2009).

Of the three transgressions identified as scientific misconduct, plagiarism seems to be declining as the use of software to detect copied text increases, Finneran noted. However, the growing number of author-pay open-access journals could increase the risk of plagiarism, because many of these journals probably are not running software to identify plagiarism.

The number of retractions has increased sharply in the past decade, and an analysis by Fang et al. (2012) of articles on PubMed found that two-thirds of retractions are due to misconduct. But retraction rates are far from a perfect measure of misconduct, Finneran cautioned. Retraction numbers are very low, and retraction has become a common practice only recently. On the other hand, many fraudulent papers are not retracted, and other proxies for misconduct have also been going up in recent years.

Reproducibility, or the inability to reproduce research results, has also been receiving more attention. For example, Ioannidis (2005) found that systematic biases led to false-positive findings in more than half of published studies, and an effort to replicate 100 psychology studies found that the mean effect size of the replications was about half of what was reported in the original articles (OSC, 2015). However, not all research is reproducible, Finneran observed. Some conditions, such as natural phenomena, cannot be replicated. Clinical trials involve individual patients, and a new cohort will differ in unspecifiable ways. When a researcher fails to reproduce someone's work, the problem could be with the reproducer rather than with the original research.

The bottom line, said Finneran, is that scientists still do not know how much misconduct is taking place. "But the important message of all of this is that we're not as good as we would like to be. We need to work harder."

*Fostering Integrity in Research* concluded that more attention needs to be focused on what it called detrimental research practices, or deviations from best practices that increase the likelihood of error and vulnerability to misconduct. These are what Finneran called "less-than-ideal, easy ways of doing science that's not quite reliable enough," including

- Detrimental authorship practices, such as honorary authorship, demanding authorship in return for access to previously collected data or materials, or denying authorship to those who deserve it;
- Not retaining or making available data, code, or other information/materials underlying research results as specified in institutional or sponsor policies, or standard practices in the field;
- Neglectful or exploitative supervision in research;
- Misleading statistical analysis that falls short of falsification;
- Inadequate institutional policies, procedures, or capacity to foster research integrity and address research misconduct allegations, and deficient implementation of policies and procedures; and
- Abusive or irresponsible publication practices by journal editors and peer reviewers.

The research community needs to be aware of these practices and address them, said Finneran. It also needs to communicate the fact that it is taking such steps to enhance its trustworthiness.

### PROMOTING TRANSPARENCY AND OPENNESS

When Marcia McNutt, the 22nd president of the NAS, was editor-in-chief of *Science*, one of her crusades, she said, was to improve practices within scientific publishing to work against violations of the norms of research and to maintain trust in science. A workshop organized by the leaders of *Science*, *Nature*, and NIH brought together the editors of top journals, the heads of major universities, and researchers to discuss reproducibility in preclinical research (McNutt, 2014). Recommendations emerging from that meeting were codified in a check sheet and have been adopted by more than 500 other journals. For example, the recommendations call for preexperimental plans for collecting and handling data, with sample size estimations to ensure appropriate signal-to-noise ratios, randomization in the treatment of samples, blind conduct of experiments, and transparency in reporting. "If one of the samples was removed and you decided, 'I'm not going to analyze this sample' and you have a good reason, you have to say so. 'I removed this sample because we found out something about this sample that was a legitimate reason to take it out of the analysis.'"

Based on the success of that workshop, a second workshop supported by the Laura and John Arnold Foundation focused on transparency as a way to address the reproducibility issue in science. That workshop led to a paper proposing what came to be known as the Transparency and Openness Promotion (TOP) guidelines, which were designed to support norms of sharing data, facilitating review and replication, and reducing bias (Nosek et al., 2015). More than 5,000 journals and organizations have agreed to the standards, making them "probably the most widely propagated standards on the planet for increasing transparency and increasing replication possibilities for science," said McNutt. The standards cover eight issues:

- Citation standards,
- Data standards,
- Analytic methods (code) transparency,
- Research materials transparency,
- Design and analysis transparency,
- Preregistration of studies,
- Preregistration of analysis plans, and
- Replication.

For each standard, journals can sign on at different levels. With the data standard, for example, the first level may simply require a journal to disclose whether data are available or, if they are not, why not. The next level may require data to be available if an article is to be published. The next level may require that the journal verify that the data have been deposited in an appropriate repository and check to see if they can be accessed. In some cases, journals may be required to access the data and ensure that they support the conclusions in the paper.

Individual disciplines may have more detailed standards. For example, a workshop on reproducibility in computational methods established the following goals (Stodden et al., 2016):

- Share data, software, workflows, and details of the computational environment in open repositories;
- Publish persistent links with a permanent identifier for data, code, and digital artifacts upon which the results depend;
- Cite shared digital scholarly objects;
- Document digital scholarly artifacts; and
- Use open licensing when publishing digital scholarly objects.

In the same issue of *Science* that contained the TOP guidelines, Alberts et al. (2015) recommended increased transparency and openness in research, the provision of more incentives for reviewing, greater recognition of excellence in reviewing, the establishment of at least two classifications of retractions to distinguish honest mistakes from fraud, and improved language in conflict-of-interest declarations. Most of these recommendations are now in the process of being instituted, McNutt reported.

Finally, McNutt et al. (2018) called for journals to adopt common and transparent standards for authorship, outline responsibilities for corresponding authors, adopt the CRediT (Contributor Roles Taxonomy) methodology for attributing contributions, and encourage authors to use the digital persistent identifier ORCID. It also recommended that institutions have regular conversations about the criteria for earning authorship on a paper.

These recommendations support those in *Fostering Integrity in Research*, which recommended that authors disclose their roles and contributions and that institutions establish policies that will thwart detrimental research practices. In addition, *Fostering Integrity in Research* urged that a Research Integrity Advisory Board be created to serve as an organizational focus for best practices and standards.

---

### The Science Behind the News: Driverless Cars

People normally think of driverless cars as devices that will make their commutes better, but "their impact is going to be so much larger than that, and it's not fully appreciated yet," said Jack Stewart, senior writer at *Wired Magazine*. When cars are continually driving around cities, picking people up and dropping them off, cities and towns will not need the parking spaces they need today. Streets and buildings can be designed differently. Accident rates will fall dramatically. Jobs driving trucks, taxis, and delivery vans will go away. The arrival of driverless cars is likely to have as great an impact on society as the arrival of smart phones, said Stewart.

The public is interested in driverless cars, as indicated by the prevalence of the phrase in online searches. But interest tends to spike when something has gone wrong, such as a crash involving an autonomous vehicle. Meanwhile, consumer awareness of autonomous vehicles remains generally low, partly because of confusing terminology, Stewart observed. Vehicles being marketed today as "self-driving," for example, are more accurately seen as cars with features to aid drivers.

He described three challenges in the news coverage of driverless cars. The first challenge is the pace of change. Companies are continually coming out with new kinds of vehicles, and "there's a lot going on behind the scenes" that typically does not make the news.

The second challenge involves morals and ethics. How safe do driverless cars need to be before they will be allowed on the roads? Twice as safe means that they would still be responsible for 15,000 to 20,000 deaths per year in the United States if all vehicular travel was done this way. "Are we going to accept that?" Stewart asked. In some hypothetical cases, driverless cars may need to kill their occupants if they are programmed to prevent a greater number of deaths otherwise. "Would you go to the car dealership and buy a car if the dealer said, 'It's got all these great features, drives itself—and, by the way, every now and then it might kill your child?'" Incorporating these kinds of moral issues into news coverage can be difficult, Stewart said.

The third challenge is the scale of the effort devoted to this technology. Many different companies, from startups to large multinationals, are currently working on driverless cars, and many are reluctant to talk about their work for competitive reasons. As a result, "keeping on top of claims is difficult."

Finally, the technology beat has its own challenges. For example, good stories need good headlines, but headlines can be too optimistic or too pessimistic, as can the overall tone of a story. "There's a balance—we have to get that right, and that's what we aim for."

Illah Nourbakhsh, professor of robotics at Carnegie Mellon University, elaborated on some of the ethical issues associated with autonomous vehicles. Complex systems that incorporate computer vision, manipulation, control, and nature are occasionally going to fail, he said. With a complex system, "you cannot possibly exhaustively test all of the parametric combinations that it will face in society," said Nourbakhsh. With millions of vehicles on U.S. roads being driven millions of miles per day, "they're going to encounter conditions that you never saw, and that's going to happen over and over again."

Also, the gap between human intelligence and autonomous devices is huge. "Your intuition about how humans fail, or what happens when you change conditions, has nothing to do with how conditions changing will cause an artificial system to behave." For example, just a few pieces of tape in the right places can fool an autonomous car into seeing a stop sign as a 45-mile-per-hour speed limit sign.

> These are not human beings. The boundary issue that we face in society ethically is that we—the policy makers, the decision makers, people in the legal profession, the insurance and reinsurance agents, and even the engineers themselves—don't understand this disparity between real technology, the promise of technology, and the reality of the boundary conditions that we will face as a society.

Peter Hancock, Provost's Distinguished Research Professor, Pegasus Professor, and Trustee Chair in the Department of Psychology and the Institute for Simulation and Training at the University of Central Florida, pointed out that cars already incorporate a tremendous amount of automation, but we do not see it. Automation does not replace human

*continued*

performance, he noted; rather, it alters human performance. For example, if people are required to respond very rarely to a situation, they will not be practiced when they need to respond.

Hancock also noted that many people like to drive, and they will be reluctant to give up their driving to a machine. For example, maybe artificial cars could teach gearheads how to drive better.

> You could take your car out on the track and do loops, and the AI will drive the first few for you, because it will do it better than a human. And then, it will gradually let you take over. It will step in where it needs to, and it will help you enjoy your car more. And then, when you've had your fun, it will deal with the stop-and-go traffic on the way home.

# 8

# Evaluating Science Communication

> **Important Points Made by the Presenter**
> - Influence on social media is a matter of the quality as well as the quantity of messages. (Fowler)
> - One way to measure the quality of a tweet is by measuring its information content. (Fowler)
> - The average information content of tweets from scientists is greater than that of tweets from nonscientists, although scientists tend to tweet less than other Twitter users. (Fowler)
> - Better measures of quality may encourage scientists to communicate more and build more of a following on social media. (Fowler)

Well-designed and rigorous evaluations can help answer critical questions in science communication. What types of narratives and images produce accurate impressions and lasting memories? How does a presentation's impact vary by audience? How should statistics and human narratives be combined for maximum impact? As the narrator of the session on evaluating science communication, Arthur Lupia, Hal R. Varian Collegiate Professor of Political Science at the University of Michigan, put it, "many researchers and science organizations have limited resources to devote to communication. Many of you are being asked to do more with less. How can we create effective science communication strategies in these circumstances?"

## INFLUENCE ON TWITTER

James Fowler, professor in the Political Science Department and in the Global Public Health Division of the Department of Medicine at the University of California, San Diego, addressed the evaluation of a specific kind of communication: tweets that involve science. Many measures exist of the influence of scientists on social media, Fowler said, including levels of activity, the responses generated by a person's activity, and a person's influence within a network of people (Riquelme and González-Cantergiani, 2016). However, none of these measures necessarily mean that a person is influential. "You could be saying things that are completely banal to millions of people and not changing their lives at all," said Fowler. "That's where a lot of these measures fall down, because they're just measuring quantity."

On Twitter, measures of quantity specifically include tweets, replies, retweets, favorites, mentions, follows, and network indicators. However, these measures have the same problem as with other social media: they do not measure influence. "It reminds me of an ancient university proverb: 'Academic deans can't read, but they sure can count,'" said Fowler. "The literature has been focusing more on what a dean would focus on."

Fowler described a way of measuring influence that incorporates quality as well as quantity. The method begins with the measure of information in a tweet. For example, if a girl gets socks from her aunt every Christmas, opening that year's present and finding socks does not provide much information. But if the present is a brand new bicycle, that girl's ideas about her aunt will change, because her aunt has done something that she has never done before.

Similarly, the quality of a tweet can be quantified by measuring how likely it is that a tweet would have occurred. According to information theory, the less probable something is, the more information it contains. Specifically, the negative logarithm of the probability can be summed across different pieces of information to get the total information in a message.

To measure the probability of a tweet, Fowler took the simple and transparent method of looking at all the words on Twitter that have occurred in the past 24 hours. He then measured all the words among responders for a 24-hour period and classified how likely it was that a given tweet would have randomly drawn from the set of all the words on

Twitter. "This is a good place to start because there will be some information in a tweet based on the kinds of words that are used" in responses to that tweet.

## SCIENTISTS ON TWITTER

Scientists tend to be in the middle of the distributions of user favorites, followers, and friends among all Twitter users, Fowler noted. However, they tend not to tweet as much. "We're busy, and tweeting isn't necessarily the first thing we do." As a result, scientists get fewer retweets per person, fewer replies per person, and fewer quotes per person. "That's not surprising. We have lower activity, so there's going to be fewer responses," Fowler said. They also get less information into the Twittersphere, because they tweet less.

However, the tweets of scientists have higher average information scores than do the tweets of nonscientists. "The moral of the story is that we don't tweet as much as other people, but when we do tweet, we tweet with a higher amount of information," he noted.

This higher information content can result in great impact, Fowler observed. As the information score of a tweet goes up, the likelihood of retweets, replies, and quotes increases. Furthermore, this increased interest can have concrete outcomes. A study of Facebook users, for example, found that messages on Facebook led about one-third of 1 million more people to vote in 2010 than would have been the case otherwise (Bond et al., 2012).

This is the kind of approach that is needed to move beyond quantity to quality, Fowler concluded. Researchers need to "try to think of creative ways where, at scale, we can measure the quality of [messages]. It's only then that we're going to be able start to think about how one scientist's actions might be able to change other people's opinions." For instance, could messages from scientists change policy or encourage people to become involved in citizen science? "I feel optimistic, given the comparisons that we're now just starting to make."

In addition, measures of quality may encourage scientists to communicate more and build more of a following, further increasing their influence. Scientists "are late comers to social media," said Fowler, "but now a lot of them have adopted it."

### Communicating Big Ideas in the Social Sciences

In the social sciences—as elsewhere in science—short and easily communicable ideas get the most attention in the media. But big social problems are harder to communicate—"they don't convey in a tweet," said Gerald Davis, Gilbert and Ruth Whitaker Professor of Business Administration at the Ross School of Business and professor of sociology at the University of Michigan.

Davis has been trying to communicate the idea that the shrinking number of corporations in the United States poses major societal challenges. The number of U.S. corporations listed on U.S. stock markets has dropped from more than 8,000 in 1997 to roughly 4,000 today. A variety of factors have contributed to this decline, including consolidation and collapse in banking, the offshoring of the electronics industry, and acquisitions in the pharmaceutical industry. Furthermore, the number of initial public offerings (IPOs) in the United States has declined substantially in the past two decades, and relatively small numbers of people work at even the most successful of these IPO companies.

Vanishing corporations pose a "giant social problem," said Davis. America's health and retirement systems were funded by corporate employers for generations. Upward mobility and the growth of the middle class depend to a major extent on stable employers with career ladders, and such employers are less common today than in the past. Large public corporations are easier to regulate than are many smaller firms. Shrinking stock markets as corporations disappear also leave fewer options for college and retirement savings. "From my perspective, what all of this means is that we are living through a foreseeable social disaster."

Davis has experimented with several ways of communicating the findings of his research. He published an academic book with a university press, but such books serve mostly to convince your colleagues, he said. He wrote a mass market book, but that only works "if you want to convince your parents," he quipped. He has been featured in segments on the Public Broadcasting System's *NewsHour* and National Public Radio, but the audiences for those programs are limited. He wrote short pieces for an Internet outlet, but this was unsatisfactory, especially when he began to read the many ill-informed and disparaging comments generated by his articles.

Davis was enthusiastic about the impact of short and accessible pieces written for reputable outlets like The Conversation (theconversation.com). The Conversation is based on academic research that addresses events of the day, with links to the underlying research. Editors work with authors to produce readable prose, and articles can be republished if attributed to The Conversation. A dashboard provides authors with data on how much, when, and where an article has been read. Such articles garner tens of thousands of readers and mostly respectful dialogue. He also encouraged scientists to respond quickly to all calls from journalists as a way of communicating their research.

His findings on the threat posed by vanishing corporations are starting to get out, Davis said, despite the communication obstacles that exist. However, powerful incentives still exist to focus on research that yields short and pithy findings. "Deans love to see it; press offices love to see Tweet-worthy work," he said. "My worry is that we're going to be moved to do much more of this Tweet-sized work in bite-sized chunks rather than real science."

# 9

# Communicating Uncertainty

> **Important Points Made by the Presenter**
>
> - Economists and other social scientists tend to make exact predictions of policy outcomes while rarely expressing uncertainty. (Manski)
> - Maintaining trust may require ways of communicating uncertainty so that false predictions do not undermine the credibility of science. (Manski)

Communicating uncertainty is one of the biggest challenges journalists face, said Laura Helmuth, national editor of health, science, and environment at *The Washington Post*, who moderated the session at the colloquium on uncertainty in science communication. Uncertainty is hard to explain and understand. Journalists typically have so much to explain in their stories that they can be tempted to leave uncertainty out. "You have to pick your explanatory battles, and this is a battle that we often put off," said Helmuth.

But journalists are getting better at it, she added. They are becoming more aware of how uncertainty can be misused, as was the case when the tobacco industry argued that the health effects of smoking were uncertain. They have learned to avoid the trap of false balance, so as not to overstate the uncertainty that exists. "We're getting better at covering uncertainty as a subject in an interesting way," said Helmuth.

## PREDICTIONS WITHOUT MEASURES OF UNCERTAINTY

Scientific evaluations of public policies should explicitly express the limits to knowledge. However, policy analysis "with what I call *incredible certitude* has been common," said Charles Manski, Board of Trustees Professor of economics in the Department of Economics at Northwestern University. The predictions that researchers make are often fragile, resting on unsupported assumptions and limited data (Manski, 2013), but economists and other social scientists tend to make exact predictions of policy outcomes while rarely expressing uncertainty. "It's not that they're fraudulent," said Manski. "It's that you assume more than you have the basis to assume using the data that you have."

Good examples, he observed, are the predictions, known as scores, made by the Congressional Budget Office (CBO) of the budgetary implications of pending federal legislation. The impacts of new legislation are difficult to foresee, yet the CBO makes 10-year point predictions, with no quantitative measure of uncertainty. Similarly, official statistics from federal agencies, such as the gross domestic product growth rate or the poverty rate, suffer from various kinds of errors, yet agencies typically report only point estimates.

Some agencies do aim to communicate uncertainty transparently. A notable case is the National Weather Service, which in a tweet issued on August 27, 2017, as rainfall from Hurricane Harvey was beginning to inundate Southeast Texas, said, "This event is unprecedented & all impacts are unknown & beyond anything experienced."

Manski listed several manifestations of incredible certitude. One is conventional certitude, which he described as statistics or estimates that are accepted as true by society but may not be true. Dueling certitudes are contradictory predictions made with alternative assumptions, as when analysts draw opposite conclusions about such issues as illegal drug policies. Conflating science and advocacy is when analysis aims to generate a predetermined conclusion, as with the practice of "model shopping," where advocates for a particular position go looking for a model that supports their views. Wishful extrapolation is using untenable assumptions to extend a conclusion in a desired direction, as when limited studies of drug outcomes are used to predict what will happen in clinical practice. Illogical certitudes draw unfounded conclusions based on deductive errors, as with research that misinterprets the heritability of personal traits. Finally, media overreach is when journalists do premature or exaggerated reporting of policy analysis.

## THE RISKS OF IGNORING UNCERTAINTY

Why do researchers express certainty when they should be expressing uncertainty? Manski pointed to two reasons. The first is that the scientific community tends to reward strong and novel findings. The second is that the public wants unequivocal policy recommendations. Analysts at the CBO, for instance, know that their point estimates should be accompanied by ranges of uncertainty. But they may believe, Manski speculated, that the members of the U.S. Congress are psychologically or cognitively unable to deal with uncertainty. (Although, he added, such estimates in the United Kingdom do include uncertainty.) Or they may believe, because the CBO has established an admirable reputation for impartiality, that it is best to leave well enough alone and have the CBO express certitude when it scores legislation, even if the certitude is conventional rather than credible.

The problem with this approach, said Manski, is that the existing social contract to take CBO scores at face value could eventually break down. Maintaining trust may require expressing uncertainty. "Once you accept incredible certitude and take numbers at face value when they shouldn't be, there may be a slippery slope from incredible certitude to utter disregard for truth. I do not think this is a second-order issue. In fact, it may be even more important to face up to uncertainty today than in the past."

---

### The Science Behind the News: Gene Drive

Gene drive refers to a technology in which the DNA of a sexually reproducing organism is altered so that most or all of its descendants, not just the typical one-half of its descendants, inherit a particular DNA segment. The technology has different methods, outcomes, and goals, said Fred Gould, University Distinguished Professor of Agriculture in the Entomology Program at North Carolina State University. It can be aimed at eradicating a species, preserving a species, or eliminating the transmission of a disease. It can be spatially or temporally restricted. It can suppress a population rather than eliminate it. However, because gene drive is a new technology, many of its applications cannot yet be foreseen.

Gene drive could also have harmful consequences, Gould observed. A DNA sequence targeted at one group of organisms could spread to other groups. Changes in an organism could alter the ecosystem of which that organism is a part. Some specific types of gene drives may be built in the future that could eliminate a group of organisms or species. The question people inevitably will ask, said Gould, is "Are you playing God?"

The day before the colloquium, two articles on gene drive were re-

*continued*

leased, one in the journal *PLoS Biology* as an opinion piece (Esvelt and Gemmell, 2017), and the other, before peer review, in the online archive *bioRxiv* (Noble et al., 2017). These two articles led to a wide variety of headlines, noted Gould, including "Genetically Engineering the Natural World, It Turns Out, Could Be a Disaster" (in *Gizmodo*), "'Gene Drives' Are Too Risky for Field Trials, Scientists Say" (in *The New York Times*), "New Zealand's War on Rats Could Change the World" (in *The Atlantic*), and "New Model Warns About CRISPR Gene Drives in the Wild" (in *Quanta*).

The abstracts of the articles themselves give a better sense of their contents, noted Gould. The abstract of the first article said that New Zealand is considering gene drives as a way to locally eliminate mammalian pests and that the article will explore the risk of accidental spread following deployment, concluding that open and international discussions are needed about a technology that could have global ramifications. However, these international discussions have been going on for nearly a decade, Gould observed, raising questions about why the researchers ignored previous work of their colleagues in raising such an alarm and why journalists amplified this alarm without checking with other sources.

The abstract of the second article noted that even the least effective gene drive systems reported to date are highly invasive and that "standard drive systems should not be developed nor field tested in regions harboring the host organism." However, the best gene drive system tested so far was not even effective at spreading in the laboratory, much less in the field, Gould noted.

How did the two papers released the day before the colloquium become such a big news story? A major reason is that gene drives are being discussed in such apocalyptic terms, Gould said. Headlines such as "Reckless Driving: Gene Drives and the End of Nature," as one article from a group of advocacy organizations was titled, raise questions about the responsibilities of researchers and professional journalists communicating about the issue. Perhaps some kind of "rules of the road" would help researchers engage with journalists in ways that would provide for more accurate reporting of scientific findings. Gould suggested that we need to ask what is the responsibility of the researcher and what is the responsibility of the journalist.

One of the authors of the first article, Kevin Esvelt, "has been doing something rather unusual in science," said Pam Belluck, a health and science writer for *The New York Times*. As part of his research into the use of gene drives to help eliminate Lyme disease from Nantucket Island, he has been allowing the community to decide whether to allow tests of the technology to occur. He has published his plans online, has made several presentations on Nantucket, and has formed steering committees that deliberately include skeptics and critics. This process has led the people of Nantucket to decide that they do not want to test gene drive, partly because it will result in bacterial genes being added to the mouse genome and partly because of possible undesirable side effects. Esvelt's approach has been praised for its outreach to the public and his willingness to change plans when the community expressed its feelings about the research.

In the perspective piece published the day before the colloquium, Esvelt and his coauthor took a much more cautious approach toward gene drive than they had in the past, writing that "now is the time to be bold in our caution." Esvelt still contends that gene drive might have applications for a plague like malaria, Belluck noted, but for now gene drive is off the table on Nantucket, even as other options are being considered to deal with Lyme disease.

Gene drive is an example of what Dominique Brossard, professor and chair of the Department of Life Sciences Communication at the University of Wisconsin–Madison, called postnormal science, where the uncertainty and stakes surrounding an issue are both high. Such issues involve not just technical risk assessments but legal, moral, social, and cultural implications. Therefore, they can generate very strong feelings in the public. The Science, Media, and the Public research group at the University of Wisconsin–Madison recently collected survey data on what Americans thought about editing genes in wildlife populations. In response to the question, "Does editing genes in wildlife population mess with nature?" more than 60 percent of respondents agreed, and 60 percent agreed that it "allows humans to play God" (Kohl et al., 2017). However, the wording of the question can make a big difference, Brossard noted. The largest response when people were asked whether "editing genes in wildlife to decrease or eliminate local populations of animals or plants that are causing environmental problems is morally acceptable" was "neither agree nor disagree," with a rough split between people who agreed with the statement and those who disagreed.

The moral context varies from place to place depending on such factors as regulation, laws, institutional rules, and religiosity, Brossard observed. In addition, the impact of a message in the media differs from place to place and among populations. For example, using genetically engineered mosquitoes as a method to reduce the threat of Zika is one circumstance under which Americans who otherwise would be reluctant to support genetic engineering may be willing to accept it as a solution to a public health problem that might personally affect them (Lull et al., 2017).

# 10

# The Role of Philanthropy in Science Communication

> **Important Points Made by the Presenters**
> - Philanthropy has a number of vital roles to advance more effective science communication among diverse stakeholders, especially around contentious issues—including helping to connect science communication research to practice. (Christopherson)
> - Improving the communication of scientists among themselves can improve their science and the science of others. (Burris)
> - As K–12 and higher education move away from formal classes that teach communication, foundations can support opportunities for scientists to learn the communication skills they will need in their professions. (Burris)
> - Philanthropic organizations can take risks and make decisions more quickly than can government, enabling these organizations to experiment with ways of supporting science communication. (Kastner)

Historically, philanthropic organizations have engaged in a wide range of activities to advance engagement with science and evidence-based policy, observed Elizabeth Christopherson, president and chief executive officer of the Rita Allen Foundation, who moderated the session on the role of philanthropy in science communication. As described in *The Oxford Handbook of the Science of Science Communication* (Jamieson et al., 2017), philanthropists and foundations have

- Promoted engagement to expand public understanding and acceptance of science;
- Conducted, commissioned, or supported research to inform policy questions;
- Catalyzed and brokered dialogue among scientists, policy makers, and other key stakeholders;
- Assessed the impact of science engagement and policy-influencing efforts; and
- Built new fields of inquiry relevant to policy formation.

The Rita Allen Foundation has supported biomedical research scientists at early stages of their careers, which provides them with greater flexibility to pursue high-risk ideas. It is also investing in innovative projects that improve the quality and accessibility of information related to science and civics. "We believe that aware, informed, and engaged citizens are vital assets for solving most critical problems in our communities," said Christopherson. The foundation has been working to connect research to practice and preparing the next generation of leaders, Christopherson said.

## PROMOTING SCIENCE COMMUNICATION AMONG SCIENTISTS

Connecting with an audience means choosing an audience with which to communicate and then understanding that audience, said John Burris, president of the Burroughs Wellcome Fund. For example, the Burroughs Wellcome Fund has recently supported the production of two whimsically illustrated middle school books that appeal both to students and to parents. Such projects, by eliciting the intellectual curiosity of students, might lead them to pursue science, to become supporters of science, or simply to understand science better, Burris observed.

In addition to supporting biomedical research, the Burroughs Wellcome Fund supports the communication of scientists among themselves, which can improve both their science and the science of others. For example, it has worked with the Howard Hughes Medical Institute on a book titled *Making the Right Moves* (Burroughs Wellcome Fund and Howard Hughes Medical Institute, 2006), which was designed in part to help scientists communicate with each other and with the public. Another way of emphasizing science communication has been to build and promote communication among its grantees. For example, the Burroughs Wellcome Fund brings in award applicants to meet with a committee to describe their results and proposals. The committee then recommends which of these applicants should be funded. "It's an expensive and time-consuming process, but in fact is extraordinarily important in our decision making," said Burris. As K–12 and higher education have moved away from speech classes and other opportunities for students to develop communication skills, future scientists have fewer opportunities to develop the skills they will need in their professions. Instead of speaking at meetings, young scientists present at poster sessions, where they speak to just one or a handful of people at a time. "It's important that you're able to get up in front of people and speak about your science," Burris said. For the same reason, the Burroughs Wellcome Fund asks its grantees to participate as much as possible in public events such as science festivals.

Philanthropies have many opportunities to try new approaches and models that in turn can be translated and scaled up, Burris noted. They can be nimbler, with less bureaucracy and fewer barriers to getting things done. Philanthropies also need to be comfortable with failure, he added. Failure can be a success if it shows that something does not work. "It's very difficult for program officers to embrace failure as a success, but in fact it is," he said.

Burris suggested experiments where science communicators try something different, followed by an evaluation to determine whether that new approach was effective. He also pointed out, however, that evaluation is "the toughest nut, in many ways, to crack." Measures need to reflect success rather than simply measuring things that are measurable. "We can count the number of publications. We can measure [the] number of grants. We can measure the number of speeches. [But] it's extraordinarily difficult to see what impact something has had."

As part of its work on communication, the Burroughs Wellcome Fund has developed an active social media presence, including accounts with Twitter, Facebook, Instagram, and Snapchat. "We have found, interestingly enough, that Twitter is far and away the communication and social media of choice. It's most obvious among science teachers. When we have a meeting, and when teachers are communicating among themselves, they use Twitter," Burris said.

## SCIENCE COMMUNICATION FOR PHILANTHROPISTS

When funding for basic research in the natural sciences began to decline over the past decade, one need that immediately became apparent was for advice to philanthropists on how to support science effectively. The Science Philanthropy Alliance was formed in 2013 to meet that need, said the organization's president, Marc Kastner. Since then, its member organizations have grown from 6 to 21, many of which have science communication as part of their missions.

A ScienceCounts survey sponsored by the Alliance members has shown that the public appreciates science but does not give it a high priority. This survey also has shown that fear, as a motivating force, does not work well with the public (though, Kastner added, fear does work well with the federal government in motivating support of basic research). With the public, hope is a more effective motivating force. For example, the widespread fascination of the public with the 2017 solar eclipse and the recent discovery of gravitational waves are examples of how scientific discoveries can excite the public.

The Alliance has commissioned what it calls "science to society stories," which describe how science has led to technologies and therapies that have improved people's lives. It has also sought to demonstrate to philanthropists the kinds of differences they can make through support of basic research. For example, "we have one philanthropist who really wants to excite the public about ocean exploration—that's his primary motivation," said Kastner.

Burris, Christopherson, and Kastner all agree that philanthropic organizations can learn from each other and then disseminate what they have learned. "The federal government is very conservative," observed Kastner. Philanthropies can take more risks and make decisions more quickly. Philanthropic organizations can also amplify potential solutions to wicked problems by working together to share best practices, collaborate on projects, and invest in research, experimentation, and evaluation, said Christopherson. As Burris said, "It's great that we as foundations are in a position to try experiments. If you give us a good one, that would be something that to me would make sense to fund."

# 11

# Reflections on the Colloquium

At the end of each day of the colloquium, two of the colloquium's organizers—Dietram Scheufele of the University of Wisconsin–Madison and Baruch Fischhoff of Carnegie Mellon University—reflected on the themes that emerged from the day's presentations and discussions. A major theme, said Scheufele, is the need for broader and more inclusive discussions about science and science communication. The application and communication of scientific results are informed by considerations that are not necessarily scientific in nature, including ethical, moral, and societal considerations. The National Academies of Sciences, Engineering, and Medicine and other organizations have done good work in convening such discussions and in studying how they should be conducted. But more research needs to be done on public discussions like consensus conferences or town halls, Scheufele said. These events tend to be attended by people who are very opposed or very supportive of a technology, whereas other communities that should be heard are often not represented.

Scheufele also pointed out that the Arthur M. Sackler Colloquia on the Science of Science Communication have been intended to galvanize a new field and a new way of thinking. The new era of science that is emerging "requires us to think differently about communication," he said. Technologies are emerging at a rapid pace and are making fast transitions from research to application. More information is available to people more quickly than ever before. People have more ability to have exchanges with others through social media. At the same time, this increased access to information has created problems, such as getting just one side of the story, or, as Scheufele noted, "we can't even make it through this conference without a whole bunch of spam coming in on the colloquium hashtag" (#SacklerSciComm).

## STORIES, REWARDS, AND RELATIONSHIPS

Fischhoff cited three themes emerging from the colloquium: one behavioral, one organizational, and one societal.

The behavioral theme is that when phenomena are complex, stories can pull diverse facts together into something that people can understand. Stories are useful if they evoke mental models, folk theories, and worldviews in ways that make sense to people, giving them "a warranted feeling of self-efficacy," said Fischhoff. "They can [then] generate appropriate conclusions from their own first principles." Science communicators can increase their effectiveness through the use of stories, but we also need "a sustained dialogue with the people we're trying to talk to, so that these are stories and issues relevant to their concerns."

The organizational theme is that academic institutions need to change their reward systems to support people who want to take a scientific approach to science communication. People need to be able to access and decode the scientific literature on science communication. They need help in evaluating their own work to determine when their intuitions about science communication might be wrong, and "we need venues for the kinds of sustained interpersonal ties, shared experiences, trust, and empathy that enable us to speak with legitimacy to our audiences."

Finally, on the societal level, it is important to provide information and establish relationships before issues polarize and spin out of control. That way, science gets a fair hearing and there is less need to blame the audience, political hysteria, or the innumeracy of the public. Scientists need help in understanding the complicated social processes through which such interactions take place, said Fischhoff.

> We need to understand when it is more important for people to express group solidarity than to endorse a fact that is absolutely at the center of our scientific life. We need to know the situations in which the facts are collateral damage to other processes. And we need to understand those situations where we're part of the problem by mixing in our preferred solutions to the problems that we're describing.

## THE POWER OF STORIES

Finally, Emmy Award–winning journalist Frank Sesno, who moderated the first day of the colloquium, elaborated on Fischhoff's point about the power of stories.

> I apologize for being so simplistic about it, but it works. A great story is compelling characters overcoming obstacles to achieve a worthy outcome. That's what science is. It's compelling characters—people in the labs, people in the field, people all over—overcoming obstacles—of the unknown, of every economic and financial sort—to achieve a worthy outcome—to gain knowledge and to move humanity forward. If we can't tell stories from science, we can't tell stories from anyplace. So there's enormous potential, up against all these challenges that we've talked about here today.

# References

Akin, H., K. M. Rose, D. A. Scheufele, M. Simis-Wilkinson, D. Brossard, M. A. Xenos, and E. A. Corley. 2017. Mapping the landscape of public attitudes on synthetic biology. *Bioscience* 67(3):290-300.

Alberts, B., R. J. Cicerone, S. E. Fienberg, A. Kamb, M. McNutt, R. M. Nerem, R. Schekman, R. Shiffrin, V. Stodden, S. Suresh, M. T. Zuber, B. K. Pope, and K. H. Jamieson. 2015. Self-correction in science at work. *Science* 348(6242):1420-1422.

ASA (American Sociological Association). 2016. *What Counts?: Evaluating Public Communication in Tenure and Promotion.* Final Report of the ASA Subcommittee on the Evaluation of Social Media and Public Communication in Sociology. Washington, DC. Available at http://www.asanet.org/sites/default/files/tf_report_what_counts_evaluating_public_communication_in_tenure_and_promotion_final_august_2016.pdf.

Bond, R. M., C. J. Fariss, J. J. Jones, A. D. I. Kramer, C. Marlow, J. Settle, and J. H. Fowler. 2012. A 62-million-person experiment in social influence and political mobilization. *Nature* 489(7415):295-298.

Burroughs Wellcome Fund and Howard Hughes Medical Institute. 2006. *Making the Right Moves: A Practical Guide to Scientific Management for Postdocs and New Faculty*, 2nd ed. Available at http://www.hhmi.org/developing-scientists/making-right-moves.

Cash, D. W., W. C. Clark, F. Alcock, N. M. Dickson, N. Eckley, D. H. Guston, J. Jaeger, and R. B. Mitchell. 2003. Knowledge systems for sustainable development. *Proceedings of the National Academy of Sciences of the United States of America* 100(14):8086-8091.

Dryden, R., M. G. Morgan, A. Bostrom, and W. Bruine de Bruin. 2017. Perceptions of how long air pollution and carbon dioxide remain in the atmosphere. *Risk Analysis* doi: 10.1111/risa.12856.

Esvelt, K. M., and N. J. Gemmell. 2017. Conservation demands safe gene drive. *PLoS Biology* 15(11):e2003850.

Fanelli, D. 2009. How many scientists fabricate and falsify research? A systematic review and meta-analysis of survey data. *PLoS ONE* 4(5):e5738.

Fang, F. C., R. G. Steen, and A. Casadevall. 2012. Misconduct accounts for the majority of retracted scientific publications. *Proceedings of the National Academy of Sciences of the United States of America* 109(42):17028-17033.

Fischhoff, B. 2013. The science of science communication. *Proceedings of the National Academy of Sciences of the United States of America* 110(Suppl 3):14031-14032.

Fischhoff, B. 2015. The realities of risk-cost-benefit analysis. *Science* 350(6260):527.

Fischhoff, B., N. Brewer, and J. S. Downs, eds. 2011. *Communicating Risks and Benefits: An Evidence-Based User's Guide*. Silver Spring, MD: U.S. Department of Health.

Fiske, S. T., and C. Dupree. 2014. Gaining trust as well as respect in communicating to motivated audiences about science topics. *Proceedings of the National Academy of Sciences of the United States of America* 111(Suppl 4):13593-13597.

Fleishman, L. A., W. Bruine de Bruin, and M. G. Morgan. 2010. Informed public preferences for electricity portfolios with CCS and other low-carbon technologies. *Risk Analysis* 30(9):1399-1410.

Guston, D. H. 2000. *Between Politics and Science: Assuring the Integrity and Productivity of Research*. New York: Cambridge University Press.

Ioannidis, J. P. A. 2005. Why most published research findings are false. *PLoS Medicine* 2(8):e124.

IOM (Institute of Medicine). 1999. *Toward Environmental Justice: Research, Education, and Health Policy Needs*. Washington, DC: National Academy Press.

IOM. 2014. *Characterizing and Communicating Uncertainty in the Assessment of Benefits and Risks of Pharmaceutical Products: Workshop Summary*. Washington, DC: The National Academies Press.

IOM and NRC (National Research Council). 2015. *Potential Risks and Benefits of Gain-of-Function Research: Summary of a Workshop*. Washington, DC: The National Academies Press.

Iyengar, S., and S. J. Westwood. 2015. Fear and loathing across party lines: New evidence on group polarization. *American Journal of Political Science* 59(3):690-707.

Iyengar, S., G. Sood, and Y. Lelkes. 2012. Affect, not ideology: A social identity perspective on polarization. *Public Opinion Quarterly* 76(3):405-431.

Jamieson, K. H., D. Kahan, and D. A. Scheufele, eds. 2017. *The Oxford Handbook of the Science of Science Communication*. New York: Oxford University Press.

John, L. K., J. G. Lowenstein, and D. Prelec. 2012. Measuring the prevalence of questionable research practices with incentives for truth telling. *Psychological Science* 23(5):524-532.

Kohl, P., D. Brossard, D. A. Scheufele, and M. A. Xenos. 2017. *Managing evolution to help nature keep pace with rapid change: Moral dimensions of proposals gene-edit wildlife*. Paper presented at the 2017 meeting of the Society for Literature, Science, and the Arts, Tempe, AZ.

Lull, R. B., D. Brossard, W. K. Hallman, and K. H. Jamieson. 2017. *The role of perceived risk of genetic engineering (GE) on public support for the release of GE mosquitoes to reduce the spread of Zika virus*. Paper presented at the Society for Risk Analysis Policy Forum, Venice, Italy.

Manski, C. 2013. *Public Policy in an Uncertain World: Analysis and Decisions*. Cambridge, MA: Harvard University Press.

Martinson, B. C., M. S. Anderson, and R. de Vries. 2005. Scientists behaving badly. *Nature* 435:737-738.

McNutt, M. 2014. Journals unite for reproducibility. *Science* 346(6210):679.

McNutt, M., M. Bradford, J. Drazen, R. B. Hanson, B. Howard, K. H. Jamieson, V. Kiermer, M. Magoulias, E. Marcus, B. K. Pope, R. Schekman, S. Swaminathan, P. Stang, and I. Verma. 2018. Transparency in authors' contributions and responsibilities to promote integrity in scientific publication. *Proceedings of the National Academy of Sciences of the United States of America* 115(11):2557-2560.

NAS (National Academy of Sciences). 2014. *The Science of Science Communication II: Summary of a Colloquium*. Washington, DC: The National Academies Press.
NAS, NAE (National Academy of Engineering), and IOM. 1992. *Ensuring the Integrity of the Scientific Research Process*. Washington, DC: National Academy Press.
NASEM (National Academies of Sciences, Engineering, and Medicine). 2015. *The Integration of Immigrants into American Society*. Washington, DC: The National Academies Press.
NASEM. 2017a. *Communicating Science Effectively: A Research Agenda*. Washington, DC: The National Academies Press.
NASEM. 2017b. *Building Communication Capacity to Counter Infectious Disease Threats: Proceedings of a Workshop*. Washington, DC: The National Academies Press.
NASEM. 2017c. *Fostering Integrity in Research*. Washington, DC: The National Academies Press.
NASEM. 2017d. *The Economic and Fiscal Consequences of Immigration*. Washington, DC: The National Academies Press.
Nicholas, G., and S. T. Fiske. In preparation. Open-ended associations to social categories.
Noble, C., B. Adlam, G. M. Church, K. M. Esvelt, and M. A. Nowak. 2017. Current CRISPR gene drive systems are likely to be highly invasive in wild populations. *bioR iv*. doi: 10.1101/219022.
Nosek, B. A., G. Alter, G. C. Banks, D. Borsboom, S. D. Bowman, S. J. Breckler, S. Buck, C. D. Chambers, G. Chin, G. Christensen, M. Contestabile, A. Dafoe, E. Eich, J. Freese, R. Glennerster, D. Goroff, D. P. Green, B. Hesse, M. Humphreys, J. Ishiyama, D. Karlan, A. Kraut, A. Lupia, P. Mabry, T. A. Madon, N. Malhotra, E. Mayo-Wilson, M. McNutt, E. Miguel, E. L. Paluck, U. Simonsohn, C. Soderberg, B. A. Spellman, J. Turitto, G. VandenBos, S. Vazire, E. J. Wagenmakers, R. Wilson, and T. Yarkoni. 2015. Promoting an open research culture: The TOP Guidelines for journals. *Science* 348(6242):1422-1425.
NRC (National Research Council). 1989. *Improving Risk Communication*. Washington, DC: National Academy Press.
NRC. 2011. *Intelligence Analysis for Tomorrow: Advances from the Behavioral and Social Sciences*. Washington, DC: The National Academies Press.
Nyhan, B., and J. Reifler. 2015. Does correcting myths about the flu vaccine work? An experimental evaluation of the effects of corrective information. *Vaccine* 33(3):459-464.
Nyhan, B., J. Reifler, S. Richey, and G. L. Freed. 2014. Effective messages in vaccine promotion: A randomized trial. *Pediatrics* 133(4):1-8.
OSC (Open Science Collaboration). 2015. Estimating the reproducibility of psychological science. *Science* 349(6251):943.
Palmgren, C. R., M. G. Morgan, W. Bruine de Bruin, and D. W. Keith. 2004. Initial public perceptions of deep geological and oceanic disposal of carbon dioxide. *Environmental Science and Technology* 38(24):6441-6450.
Rhoten, D. 2003. *Final Report: A Multi-Method Analysis of the Social and Technical Conditions for Interdisciplinary Collaboration*. San Francisco, CA: The Hybrid Vigor Institute.
Riquelme, F., and P. González-Cantergiani. 2016. Measuring user influence on Twitter: A survey. *Information Processing and Management* 52(5):949-975.
Scheufele, D. A., M. A. Xenos, E. L. Howell, K. M. Rose, D. Brossard, and B. W. Hardy. 2017. U.S. attitudes on human genome editing. *Science* 357(6351):553-554.
Smith, M. J., S. S. Ellenberg, L. M. Bell, and D. M. Rubin. 2008. Media coverage of the measles-mumps-rubella vaccine and autism controversy and its relationship to MMR immunization rates in the United States. *Pediatrics* 121:e836-e843.
Stodden, V., M. McNutt, D. H. Bailey, E. Deelman, Y. Gil, B. Hanson, M. A. Heroux, J. P. A. Ioannidis, and M. Taufer. 2016. Enhancing reproducibility for computational methods. *Science* 354(6317):1240-1241.

# Appendix A

# Agenda

---

**Thursday, November 16, 2017**

Emcee: Frank Sesno (The George Washington University)

8:30–8:45　　*Welcome*
　　　　　　Marcia McNutt, President, National Academy of Sciences

8:45–9:30　　*Future Directions in the Sciences of Science Communication: A Discussion of the NASEM Report* **Communicating Science Effectively: A Research Agenda**
　　　　　　Alan Leshner (American Association for the Advancement of Science, Emeritus)

　　　　　　Discussant: Baruch Fischhoff (Carnegie Mellon University)

9:30–10:10　 *A View from Philanthropy on the Future of Science Communication*
　　　　　　Session Moderator: Elizabeth Christopherson (Rita Allen Foundation)

　　　　　　Discussion with John Burris (Burroughs Wellcome Fund) and Marc Kastner (Science Philanthropy Alliance)

10:10–10:40　Break

10:40–12:00  *Creating a Collaborative Community for the Sciences of Science Communication*
Gerald Davis (University of Michigan) and Laurie Weingart (Carnegie Mellon University)

Discussants: David Guston (Arizona State University) and Kathleen Tierney (University of Colorado Boulder)

12:00–1:15  Lunch

1:15–2:45  *Marshalling the Troops: How Can Traditional Disciplines Help Build the Scale of Research in Science Communication?*
Session Moderator: Ken Prewitt (Columbia University)

*Communicating with the Public About Energy*
Wändi Bruine de Bruin (University of Leeds) and Granger Morgan (Carnegie Mellon University)

*Communicating with the Public About Research on Immigration*
Shanto Iyengar (Stanford University) and Doug Massey (Princeton University)

*Communicating with the Public About Infectious Disease*
Bob Hornik (University of Pennsylvania) and Susan Scrimshaw (Nevin Scrimshaw International Nutrition Foundation)

2:45–3:15  Break

3:15–3:55  *Building Capacity for Science Communication Partnership Award 1: Evidence-Based Science Communication to Policy Makers*
Elizabeth Suhay (American University), Emily Cloyd (American Association for the Advancement of Science), and Erin Nash (Durham University)

Discussants: Jim Cohen (The Kavli Foundation), Fay Cook (National Science Foundation), and David Herring (National Oceanic and Atmospheric Administration)

3:55–4:55     *Science in the News: Human Genome Editing*
              Cornelia Dean (*The New York Times*), Matthew Porteus
              (Stanford University), and Dietram Scheufele (University
              of Wisconsin–Madison)

4:55–5:00     *Wrap-Up/Lessons Learned*
              Baruch Fischhoff (Carnegie Mellon University)

5:00–6:15     *Reception*

6:15–7:30     *The Mistrust of Science*
              Introduction: Alan Leshner (American Association for the
              Advancement of Science, Emeritus)

              Atul Gawande (Brigham and Women's Hospital)

## Friday, November 17, 2017

Emcee: Ashley Llorens (John Hopkins Applied Physics Lab)

7:15–8:30     *Invitation Only Collaboration Breakfast for Leaders of
              Philanthropic Organizations*

8:40–8:45     *Welcome*
              Alan Leshner (American Association for the Advancement
              of Science, Emeritus)

8:45–9:20     *Rethinking Evaluation: Using Networks, Big Data, and
              Social Media to Measure Dissemination and Impact*
              Session Moderator: Arthur Lupia (University of Michigan)

              James Fowler (University of California, San Diego)

9:20–10:20    *Science in the News: Artificial Intelligence and Driverless
              Cars*
              Jack Stewart (*Wired*), Illah Nourbakhsh (Carnegie Mellon
              University), and Peter Hancock (MIT2 Laboratory;
              University of Central Florida)

10:20–10:45   Break

10:45–12:00 **Incentives for Scientists and Engineers to Communicate About Their Research: Roundtable Discussion**
Session Moderator: Andrew Hoffman (University of Michigan)

Discussion with Neil Donahue (Carnegie Mellon University), KerryAnn O'Meara (University of Maryland, College Park), Dietram Scheufele (University of Wisconsin–Madison), Ahna Skop (University of Wisconsin–Madison), and Emmanuel Taylor (Energetics Incorporated)

12:00–1:00 Lunch

1:00–1:40 **Building Capacity for Science Communication Partnership Award 2: Evaluating New Approaches to Promoting Vaccination**
Brendan Nyhan (Dartmouth College) and Christine Finley (Vermont Department of Health)

Discussants: Greg Boustead (Simons Foundation), Suzanne Ffolkes (Research!America), Paul Hanle (Climate Central), and Doron Weber (Alfred P. Sloan Foundation)

1:40–3:05 **Focus on a Communication Challenge: Threats to Science's Reputation**
Session Moderator: Kathleen Hall Jamieson (University of Pennsylvania)

*Reputation: What's at Stake?*
Susan Fiske (Princeton University)

*What Is the Extent of the Problem?*
Kevin Finneran (*Issues in Science and Technology*)

*Threats to Science: Exploring Solutions*
Marcia McNutt (National Academy of Sciences)

3:05–3:35 Break

3:35–4:10   *The Role of Scientists and the Media in Communicating Uncertainty*
Session Moderator: Laura Helmuth (*The Washington Post*)

Charles Manski (Northwestern University)

4:10–5:10   *Science in the News: Gene Drive*
Fred Gould (North Carolina State University), Pam Belluck (*The New York Times*), and Dominque Brossard (University of Wisconsin–Madison)

5:10–5:15   *Concluding Remarks*
Dietram Scheufele (University of Wisconsin–Madison)

# Appendix B

# Speakers

**Maria Balinska** is the editor of The Conversation US and founder of Latitude News. Balinska is an award-winning American journalist with more than 10 years of experience in senior management at the British Broadcasting Corporation (BBC) in London. As the editor of World Current Affairs Radio, she led the team producing specialist international content designed to complement the daily news agenda and attract new audiences to international affairs. During her tenure as editor, Balinska launched and executive produced nine new programs for the BBC, including *Crossing Continents*, "one of the BBC's most reliable current affairs programs" (*The Guardian*) and, most recently, BBC Radio's weekly magazine show about the United States, *Americana*.

A graduate of Princeton University and the University of Maryland's School of Public Policy, and a 2010 Nieman Fellow at Harvard, Balinska is also the author of *The Bagel: The Surprising History of a Modest Bread*, a book described by *Slate* as "lively and well researched" and by *The New York Times* as "scrumptious."

**Pam Belluck** is an American journalist and an author, a health and science writer for *The New York Times*, and the author of the acclaimed nonfiction book *Island Practice*. In 2015, she was a member of *The New York Times* reporting team that received a Pulitzer Prize for its coverage of the Ebola epidemic. Her honors include a Fulbright Scholarship and a Knight Journalism Fellowship. Her work has been chosen for *The Best American Science Writing*.

Belluck was selected to be the Ferris Professor of Journalism at Princeton in 2014 and has taught and spoken about science journalism in various venues, including the Santa Fe Science Writing Workshop, the Simons Foundation, the American Association for the Advancement of Science convention, and on *The New York Times* Journeys voyage to the Galapagos Islands. She has appeared on numerous radio and television news shows, is a member of the TEDMED editorial advisory board, and served on a journalism advisory committee for the American Academy of Arts & Sciences. *Island Practice*, a true tale about a colorful, contrarian doctor on Nantucket, has been optioned for a television series.

**Greg Boustead** joined the Simons Foundation in 2012 as community manager for *Spectrum*. In 2015, he moved to the foundation's Education & Outreach division, where he helped launch Science Sandbox, a new initiative dedicated to inspiring a deeper interest in science, especially among those who do not think of themselves as science enthusiasts. Before joining the foundation, he was editorial producer for the World Science Festival, researching and producing its public science programs. Previously, Boustead was senior editor of the science and culture magazine *Seed*, and he has contributed as a freelancer to *Vice*, *Motherboard*, and *Scientific American*. He has a B.S. in psychology and a B.A. in English from the University of Florida.

**Dominique Brossard** is a professor in and the chair of the Department of Life Sciences Communication at the University of Wisconsin–Madison. Dr. Brossard is also an affiliate of the university's Robert F. & Jean E. Holtz Center for Science and Technology Studies and Institute for Regional and International Studies, as well as the Morgridge Institute for Research. Dr. Brossard is known for her research in the field of science communication, specializing in the impact of new media environments and public opinion dynamics about contested issues in science.

A fellow of the American Association for the Advancement of Science and a former board member of the International Network of Public Communication of Science and Technology, Dr. Brossard is an internationally known expert in public opinion dynamics related to controversial scientific issues. She has served on various committees producing reports for the National Academies of Sciences, Engineering, and Medicine, including the most recent comprehensive report on genetically engineered crops, *Genetically Engineered Crops: Experiences and Prospects*. Dr. Brossard earned her M.S. in plant biotechnology from the École Nationale d'Agronomique de Toulouse and her M.P.S. and Ph.D. in communication from Cornell University.

**Wändi Bruine de Bruin** is the university leadership chair in behavioral decision making at the Leeds University Business School, where she also serves as co-director of the Centre for Decision Research. She holds affiliations with Carnegie Mellon University, the University of Southern California, and the RAND Corporation. Her research focuses on behavioral decision making, individual differences in decision-making competence across the life span, and risk perception and communication. Dr. Bruine de Bruin is a member of the editorial boards of the *Journal of Risk Research*, the *Journal of Experimental Psychology: Applied*, the *Journal of Behavioral Decision Making, Medical Decision Making*, and *Psychology and Aging*. She is a member of the Scientific & Technical Committee of the International Risk Governance Council, which provides evidence-based advice to international policy makers. She has contributed her expertise to numerous expert panels and committees, including the National Academies of Sciences, Engineering, and Medicine's Committee on the Science of Science Communication: A Research Agenda. Dr. Bruine de Bruin received a B.Sc. and an M.Sc. in psychology and cognitive psychology, respectively, from Free University Amsterdam and an M.Sc. and a Ph.D. in behavioral decision theory and psychology from Carnegie Mellon University.

**John Burris** became the president of the Burroughs Wellcome Fund in July 2008. Dr. Burris is the former president of Beloit College. He has served as the president of the American Institute of Biological Sciences and is or has been a member of a number of distinguished scientific boards and advisory committees, including the Grass Foundation; the Stazione Zoologica "Anton Dohrn" in Naples, Italy; the American Association for the Advancement of Science; and the Radiation Effects Research Foundation in Hiroshima, Japan. He has also served as a consultant to the National Conference of Catholic Bishops' Committee on Science and Human Values. From 1984 to 1992, Dr. Burris served as the executive director of the National Research Council's Commission on Life Sciences. Prior to his appointment at Beloit in 2000, Dr. Burris served for 8 years as the director and chief executive officer of the Marine Biological Laboratory in Woods Hole, Massachusetts. Dr. Burris received an A.B. in biology from Harvard University in 1971, attended the University of Wisconsin–Madison in an M.D.–Ph.D. program, and received a Ph.D. in marine biology from the Scripps Institution of Oceanography at the University of California, San Diego, in 1976.

**Elizabeth Christopherson** is the president and chief executive officer of the Rita Allen Foundation, an organization that invests in transformative ideas in their earliest stages. Christopherson joined the foundation as a trustee in 2009, and she is now guiding it through a period of rapid expansion into new funding areas, including engagement, civic literacy, community building, and leadership in science and social innovation. She has served on many regional and national boards, including as president of the New Jersey Women's Forum. Christopherson is a recipient of five honorary degrees and numerous awards for public service, including the International Women's Forum Women Who Make a Difference Award. An advocate for the arts, Christopherson is the former chair of the New Jersey State Council on the Arts, where she led the creation of two arts plans for the state. She was the first female executive director of New Jersey's public broadcasting network. Christopherson is a graduate of Wellesley College, where she studied Chinese language, history, and society.

**Emily Cloyd** is the project director for public engagement at the American Association for the Advancement of Science (AAAS), where she leads day-to-day operations of the AAAS Center for Public Engagement with Science and Technology. Cloyd is an expert in science communication and public engagement and has a particular interest in the use of environmental science to support decision making. Prior to joining AAAS, she led engagement and outreach efforts at the U.S. Global Change Research Program. Cloyd holds a Master of Professional Studies from the State University of New York College of Environmental Science and Forestry and a B.S. in plant biology from the University of Michigan.

**James Cohen** is the director of communications and public outreach for The Kavli Foundation, which is based in Los Angeles, California, and is dedicated to advancing science for the benefit of humanity, promoting public understanding of scientific research, and supporting scientists and their work. As director, Cohen is on a team that provides strategic direction and oversight for the foundation's communications initiatives and programs, from the support of science journalism to helping scientists become better communicators, along with targeted direct public outreach activities. Prior to joining the foundation, Cohen was the director of media relations as well as associate director of communications at the University of California, Irvine, where he served during the tenures of chancellors Ralph Cicerone and Michael Drake. A native of New York City, he is a member of The Authors Guild and the Writers Guild of America, West, and a graduate of the Columbia University Graduate School of Journalism.

**Fay Lomax Cook** is the assistant director for the Directorate for Social, Behavioral and Economic Sciences at the National Science Foundation. Dr. Cook is a professor at Northwestern University, where she is a faculty fellow of the Institute for Policy Research and a professor of human development and social policy. She is a leading social science researcher. Her research focuses on the interrelations between public opinion and social policy, the politics of public policy, public deliberation, energy policy, and support for programs for older Americans, particularly Social Security.

Dr. Cook has written and published numerous scholarly articles and books. She has held many noteworthy national and international positions, including the president of the Gerontological Society of America, a fellow at the Center for Advanced Study in the Behavioral Sciences, and a visiting scholar at the Russell Sage Foundation. She is a fellow of the Gerontological Society of America and an elected member of the National Academy of Social Insurance. Dr. Cook received her M.A. and Ph.D. in social policy from The University of Chicago.

**Karen Cook** is the Ray Lyman Wilbur Professor of Sociology, the director of the Institute for Research in the Social Sciences, and the vice-provost for faculty development and diversity at Stanford University. Dr. Cook conducts research on social exchange networks, power and influence dynamics, intergroup relations, negotiation strategies, social justice, and trust in social relations. Her research underscores the importance of trust in facilitating exchange relationships and of networks in creating social capital—for example, in physician–patient interactions and their effect on health outcomes. Dr. Cook has edited and co-edited a number of books in the Russell Sage Foundation Trust Series, is co-author of *Cooperation Without Trust?*, and co-edited *Sociological Perspectives on Social Psychology*. In 1996, she was elected to the American Academy of Arts & Sciences and in 2007 to the National Academy of Sciences. In 2004, she received the Cooley Mead Award of the American Sociological Association's Social Psychology Section for career contributions to social psychology. Dr. Cook received her M.A. and Ph.D. in sociology from Stanford University.

**Gerald (Jerry) Davis** is an American sociologist and the Gilbert and Ruth Whitaker Professor of Business Administration at the Ross School of Business and a professor of sociology at the University of Michigan. Dr. Davis is known for his research on corporate networks, social movements, and organization theory. His research is broadly concerned with the effects of finance on society. His most recent research examines how ideas about corporate social responsibility have evolved to meet changes in the structures and geographic footprint of multinational corporations, whether "shareholder capitalism" is still a viable model for economic development, and how income inequality in an economy is related to corporate size and structure.

Dr. Davis's books include *Social Movements and Organization Theory* and *Changing Your Company from the Inside Out: A Guide for Social Intrapreneurs*. He has published widely in management, sociology, and finance. He is currently the editor of the *Administrative Science Quarterly* and director of the Interdisciplinary Committee on Organization Studies at the University of Michigan. Dr. Davis received his Ph.D. in organizational behavior from the Graduate School of Business at Stanford University.

**Cornelia Dean** is a science writer and the former science editor of *The New York Times* and the Distinguished Visiting Lecturer in Environmental Studies at Brown University. As the science editor, Dean was responsible for the coverage of science, engineering, health, and medicine news in both the daily paper and in the weekly science section. She is the author of *Making Sense of Science: Separating Substance from Spin* and *Am I Making Myself Clear?: A Scientist's Guide to Talking to the Public*, both published by Harvard University Press. Before her appointment by Brown, Dean taught undergraduate and graduate seminars on the public's understanding of science, environmental policy, and other issues at Harvard, where she was twice honored for distinction in teaching. She is a fellow of the American Association for the Advancement of Science. Dean received her M.S. in journalism from Brown University.

**Neil Donahue** is the Lord Professor of Chemistry in the Departments of Chemistry, Chemical Engineering, and Engineering and Public Policy and director of the Steinbrenner Institute for Environmental Education and Research at Carnegie Mellon University. His research group focuses on the behavior of organic compounds in Earth's atmosphere. They are world-renowned experts in studying what happens to compounds from both natural sources and human activity when they are emitted into the atmosphere. He is a research team member of the CLOUD experiment at CERN exploring atmospheric new-particle formation. This research has led to three publications in *Nature*, two in *Science*, and three in the *Proceedings of the National Academy of Sciences* (*PNAS*) over the past 5 years. He also is an author of four other publications in *PNAS* in the past 5 years. He is a fellow of the American Geophysical Union and has won the Pittsburgh and Esselen Awards from the American Chemical Society, as well as the Carnegie Science Award for the Environment. Recently his research has focused on the origin and transformations of very small organic particles, which play a critical role in climate change and human health. Dr. Donahue received an A.B. in physics from Brown University in 1985, a Ph.D. in meteorology from Massachusetts Institute of Technology in 1991, and spent 9 years as a research scientist at Harvard University before returning to Pittsburgh in 2000.

**Suzanne Ffolkes** is the vice president of communications at Research!America, the nation's largest nonprofit advocacy alliance working to make research to improve health a higher national priority. As the director of media advocacy at the American Heart Association (AHA), Ffolkes oversaw strategic media advocacy campaigns at the federal, state, and local levels. She significantly expanded the scope of the organization's media advocacy campaigns to raise awareness and build support for the association's policy agenda. Prior to joining the AHA, Ffolkes held key communications and media positions with the American Federation of Labor, Congress of Industrial Organizations, and United Negro College Fund. Preceding her work with organizations in the nonprofit sector, she spent nearly 12 years as a broadcast news editor, radio anchor, and reporter in Houston, Texas; Wilmington, Delaware; and at the Associated Press Broadcast News Center in Washington, DC. Ffolkes received an M.A. in public communications and a B.A. in broadcast journalism and communications from American University.

**Christine Finley** has experience in public heath, clinical practice, and education. She has held various leadership roles in the Vermont Department of Health, and has been the immunization manager for the past 7 years. She has conducted immunization research with the Centers for Disease Control and Prevention, and Vermont Department of Health staff. She currently serves as a liaison to the Advisory Committee on Immunization Practices for the Association of Immunization Managers. She has worked as a nurse practitioner in Vermont and abroad.

**Kevin Finneran** has been the editor-in-chief of *Issues in Science and Technology* since 1991. Prior to that, he was the Washington editor of *High Technology* magazine, a correspondent for the *London Financial Times* energy newsletters, and a consultant on science and technology policy. His clients included the National Science Foundation, the Office of Technology Assessment, the U.S. Agency for International Development, and the U.S. Environmental Protection Agency. Prior to launching his career in science and technology policy, he taught literature and film studies at Rutgers University. He is a fellow of the American Association for the Advancement of Science and the author of *The Federal Role in Research and Development: Report of a Workshop* (National Academy Press, 1986) and a contributing author to *Future R&D Environments: A Report for the National Institute of Standards and Technology* (National Academy Press, 2002).

**Baruch Fischhoff** is the Howard Heinz University Professor at the Institute for Politics and Strategy and the Department of Engineering and Public Policy at Carnegie Mellon University. Dr. Fischhoff has expertise in decision making and risk analysis and works with students studying the decision sciences. He is a member of the National Academy of

Medicine and the National Academy of Sciences and is a past president of the Society for Judgment and Decision Making and the Society for Risk Analysis. He also has been a member of the Eugene, Oregon, Commission on the Rights of Women; the U.S. Department of Homeland Security's Science and Technology Advisory Committee; and the U.S. Environmental Protection Agency's Scientific Advisory Board, where he chaired the Homeland Security Advisory Committee. A graduate of Detroit public schools, Dr. Fischhoff holds a B.S. in mathematics and psychology from Wayne State University and an M.A. and a Ph.D. in psychology from the Hebrew University of Jerusalem.

**Susan Fiske** is the Eugene Higgins Professor of Psychology and Public Affairs at Princeton University and currently chairs the National Academies of Sciences, Engineering, and Medicine's Board on Behavioral, Cognitive, and Sensory Sciences. Dr. Fiske was elected to the National Academy of Sciences in 2013 and is a member of the National Academy of Sciences Council. She has published more than 350 articles on social cognition and investigates cognitive stereotypes and emotional prejudices at cultural, interpersonal, and neural levels. Dr. Fiske's most recent book is *The Human Brand: How We Respond to People, Products, and Companies*, co-authored with Chris Malone. With Shelley Taylor, she has written five editions of the graduate text *Social Cognition* and as sole author, three editions of the advanced undergraduate text *Social Beings: Core Motives in Social Psychology*. She is the editor of the *Annual Review of Psychology*, on the editorial board of the *Proceedings of the National Academy of Sciences*, and is the founder and editor of *Policy Insights from Behavioral and Brain Sciences*. Dr. Fiske received a B.A. in social relations and a Ph.D. in social psychology from Harvard University.

**James Fowler** is a professor in the Political Science Department and in the Global Public Health Division of the Department of Medicine at the University of California, San Diego. His work lies at the intersection of the natural and social sciences, with a focus on social networks, behavior, evolution, politics, genetics, and big data. Dr. Fowler has been named a fellow of the John Simon Guggenheim Foundation; one of *Foreign Policy*'s Top 100 Global Thinkers; *TechCrunch*'s Top 20 Most Innovative People; *Politico*'s 50 Key Thinkers, Doers, and Dreamers; and Most Original Thinker of the year by The McLaughlin Group. His research has been featured in numerous best-of lists, including *The New York Times Magazine*'s Year in Ideas, *Time Magazine*'s Year in Medicine, *Discover Magazine*'s Year in Science, and *Harvard Business Review*'s Breakthrough Business Ideas.

    Together with Nicholas Christakis, Dr. Fowler wrote *Connected: The Surprising Power of Social Networks and How They Shape Our Lives*. Winner of a Books for a Better Life Award, *Connected* has been translated into 20 languages, named an Editor's Choice by *The New York Times* Book Review, and featured in *Wired Magazine*, Oprah's Reading Guide, *Business Week*'s Best Books of the Year, *GOOD Magazine*'s 15 Books You Must Read, and a cover story in *The New York Times Magazine*. Dr. Fowler earned his Ph.D. and M.A. in government from Harvard University and his M.A. in international relations from Yale University.

**Atul Gawande** is a surgeon, writer, and public health researcher. Dr. Gawande practices general and endocrine surgery at Brigham and Women's Hospital. He is a professor in the Department of Health Policy and Management at the Harvard T.H. Chan School of Public Health and the Samuel O. Thier Professor of Surgery at Harvard Medical School. He is also executive director of Ariadne Labs, a joint center for health systems innovation, and chairman of Lifebox, a nonprofit organization making surgery safer globally.

Dr. Gawande has been a staff writer for *The New Yorker* magazine since 1998 and has written four *New York Times* best-sellers: *Complications*, *Better*, *The Checklist Manifesto*, and most recently, *Being Mortal: Medicine and What Matters in the End*. He is the winner of two National Magazine Awards, AcademyHealth's Impact Award for highest research impact on health care, a MacArthur Fellowship, and the Lewis Thomas Award for writing about science. Dr. Gawande received his M.D. from Harvard Medical School and completed his residency in general surgery at Brigham and Women's Hospital. He is an elected member of the National Academy of Medicine.

**Fred Gould** is the William Neal Reynolds Professor of Agriculture in the Entomology Program at North Carolina State University (NC State). Dr. Gould conducts cutting-edge research in the areas of ecology and evolutionary biology and studies the ecology and genetics of insect pests to improve food production and human and environmental health. He is an elected member of the National Academy of Sciences. In 2012, Dr. Gould was the 10th faculty member from NC State to win the O. Max Gardner Award. From 2013 to 2015, he served on the National Academies of Sciences, Engineering, and Medicine's Roundtable on Public Interfaces of the Life Sciences. Dr. Gould has participated in policy development for transgenic crops at the national and international levels. He has authored more than 160 refereed publications and has been invited to present papers at numerous conferences, symposia, and seminars. Dr. Gould received a B.S. in biology from Queens College and a Ph.D. in ecology and evolutionary biology from the State University of New York at Stony Brook.

**David Guston** is a professor in and the founding director of the School for the Future of Innovation in Society at Arizona State University (ASU), where he is also co-director of the Consortium for Science, Policy & Outcomes. Additionally, Dr. Guston is a principal investigator with and the director of the Center for Nanotechnology in Society at ASU, a National Science Foundation (NSF)-funded Nanoscale Science and Engineering Center dedicated to studying the societal aspects of nanoscale science and engineering research and improving the societal outcomes of nanotechnologies through enhancing the societal capacity to understand and make informed choices. Dr. Guston is widely published and cited on research and development policy, technology assessment, public participation in science and technology, and the politics of science policy. He is currently the founding editor of the *Journal of Responsible Innovation*. He has also served on the NSF's review panel on Societal Dimension of Engineering, Science, Technology, and on the National Academy of Engineering's steering committee on Engineering Ethics and Society. Dr. Guston received a B.A. from Yale University and a Ph.D. from the Massachusetts Institute of Technology.

**Peter Hancock** is the Provost's Distinguished Research Professor, Pegasus Professor, and Trustee Chair in the Department of Psychology and the Institute for Simulation and Training at the University of Central Florida. Additionally, Dr. Hancock is a director of the Minds in Technology, Machines in Thought (MIT2) research laboratories in the Department of Civil and Environmental Engineering. Prior to his current position, he founded and was the director of the Human Factors Research Laboratory at the University of Minnesota, where he held appointments as a professor in computer science and electrical engineering, mechanical engineering, psychology, and kinesiology as well as in the Cognitive Science Center and the Center on Aging Research. Dr. Hancock was principal investigator on the Multi-Disciplinary University Research Initiative, in which he oversaw $5 million of funded research on stress, workload, and performance. His current experimental work concerns the evaluation of behavioral response to high-stress conditions. His theoretical works concern human relations with technology and the possible futures of this symbiosis.

**Paul Hanle** was elected the president and the chief executive officer of Climate Central in April 2011. From 2000 to 2011, Dr. Hanle was the president of the Biotechnology Institute, an independent nonprofit organization dedicated to biotechnology education. As its first president, he built the Institute into the leading national organization in its field. Prior to serving as the president of the Biotechnology Institute, Dr. Hanle was the executive director of the Maryland Science Center, Baltimore's hands-on science museum, from 1987 until 1996. He then became the president of the Academy of Natural Sciences of Philadelphia, the nation's oldest museum of natural history and a leading center of environmental research. He earned a Ph.D. in the history of science and medicine and an M.S. in physics from Yale University. He received his A.B. in physics from Princeton University in 1969. He was a member of the Institute for Advanced Study in Princeton during the 1983–1984 academic year.

**Laura Helmuth** is *The Washington Post*'s national editor of health, science, and environment. Prior to her current role, Dr. Helmuth served as the director of digital news at *National Geographic* and as the science and health editor at *Slate* magazine, where she was responsible for *Slate*'s imaginative science coverage, including a fascinating series on the doubling of the human life span that asked readers to share why they are not dead yet. Dr. Helmuth has been a health and science editor for nearly two decades, including at *Smithsonian* and *Science* magazines. She is the president of the National Association of Science Writers. Dr. Helmuth holds a Ph.D. in cognitive neuroscience from the University of California, Berkeley. She also has a large following on Twitter at @laurahelmuth, where she delivers wide-ranging curation of and commentary on the latest health and science news.

**David Herring** is a science writer and an editor with extensive experience communicating about climate and earth science. In March 2008, Herring joined the National Oceanic and Atmospheric Administration's (NOAA's) Climate Program Office, where he serves as the program manager of the Communication, Education, and Engagement Division. He is also the program manager of NOAA's Climate.gov and leads the Climate Literacy Objective for NOAA's Climate Mission Goal. Before coming to NOAA, Herring worked for 16 years in the Earth Sciences Division at NASA's Goddard Space Flight Center, where he served as project manager for education and outreach, team leader for NASA's Earth Observatory, and outreach coordinator for the Terra satellite mission.

Herring trained in journalism, science education, and science and technical communication at East Carolina University, in Greenville, North Carolina, where he received his M.A. in 1992. He is an elected fellow of the American Association for the Advancement of Science.

**Andy Hoffman** is the Holcim (US), Inc., Professor of Sustainable Enterprise at the University of Michigan, a position that holds joint appointments in the Stephen M. Ross School of Business and the School for Environment and Sustainability. Dr. Hoffman's research uses organizational behavior models and theories to understand the cultural and institutional aspects of environmental issues for organizations. He has published more than 100 articles and book chapters, as well as 14 books, which have been translated into 5 languages. He also writes about the role of academic scholars in public and political discourse. He has been awarded the Aspen Institute Faculty Pioneer Award (2016), American Chemical Society National Award (2016), and Strategic Organization Best Essay Award (2016).

His work has been covered in numerous media outlets, including *The New York Times*, *Time*, *The Wall Street Journal*, and *National Geographic*. He has served on numerous research committees for the National Academies of Sciences, Engineering, and Medicine; the Johnson Foundation; the Climate Group; the China Council for International Cooperation on Environment and Development; and the Environmental Defense Fund. Dr. Hoffman serves on advisory boards for ecoAmerica, Next Era Renewable Energy Trust, SustainAbility, the Michigan League of Conservation Voters, the Center for Environmental Innovation, and the Stanford Social Innovation Review.

**Robert Hornik** is the Wilbur Schramm Professor of Communication and Health Policy at The Annenberg School for Communication at the University of Pennsylvania. Since 2013, he has been the co-director of the Penn Tobacco Center of Regulatory Science, a first-of-its-kind regulatory science research enterprise aimed at informing the regulation of tobacco products to protect public health. Dr. Hornik led the Center of Excellence in Cancer Communication Research at the University of Pennsylvania from 2003 to 2014. His most recent research focuses on how Americans are affected by their exposure to information about cancer prevention, screening, and treatment; the effects of new and old media content on tobacco-related beliefs and behavior among youth and young adults; and the development and validation of methods for choosing preferred message themes for communication campaigns. Dr. Hornik has particular expertise in research methods for determining the effects of public health communication interventions and of media exposure. He received an A.B. in international relations from Dartmouth College and an M.A. and a Ph.D. in communication research from Stanford University.

**Shanto Iyengar** holds the Chandler Chair in Communication at Stanford University, where he is also a professor of political science and director of the Political Communication Laboratory. Dr. Iyengar's areas of expertise include the role of mass media in democratic societies, public opinion, and political psychology. His research has been supported by grants from the National Science Foundation, the National Institutes of Health, the Ford Foundation, The Pew Charitable Trusts, and the Hewlett Foundation. He is the recipient of several professional awards, including the Philip

Converse Award of the American Political Science Association for the best book in the field of public opinion, the Murray Edelman Lifetime Achievement Award, and the Goldsmith Book Prize from Harvard University. Dr. Iyengar is author or co-author of several books, including *Is Anyone Responsible?* and *News That Matters*. He received his B.A. from Linfield College and his Ph.D. from The University of Iowa.

**Kathleen Hall Jamieson** is the Elizabeth Ware Packard Professor of Communication at The Annenberg School for Communication and the Walter and Leonore Annenberg Director of the Annenberg Public Policy Center at the University of Pennsylvania. She helped create FactCheck.org and FlackCheck.org, two nonpartisan projects of the Annenberg Public Policy Center that monitor deception in U.S. politics. Dr. Jamieson is a fellow of the American Academy of Arts & Sciences, the American Philosophical Society, the American Academy of Political and Social Science, and the International Communication Association. She has won university-wide teaching awards at each of the three universities where she has taught and political science or communication awards for four of her books. Dr. Jamieson received a B.A. in rhetoric and public address from Marquette University and her M.A. and Ph.D. in communications arts from the University of Wisconsin–Madison.

**Marc Kastner** is the president of the Science Philanthropy Alliance, a coalition of leading nonprofit institutions and foundations dedicated to increasing financial support for basic science research. Prior to his current role, Dr. Kastner had a long career in a variety of senior positions at the Massachusetts Institute of Technology. He was named the Donner Professor of Physics in 1989, and became the director of the Center for Materials Science and Engineering in 1993, head of the Department of Physics in 1998, and dean of the School of Science in 2007. Dr. Kastner has served as chair of the Solid State Sciences Committee and as chair of the Board of Physics and Astronomy of the National Academies of Sciences, Engineering, and Medicine. He also served on the Science Advisory Boards of the National Cancer Institute and the Gordon and Betty Moore Foundation. In 1995, he received the David Adler Lectureship Award of the American Physical Society. Dr. Kastner is a member of the National Academy of Sciences, a fellow of the American Academy of Arts & Sciences, the American Physical Society, and the American Association for the Advancement of Science. Dr. Kastner received his Ph.D. and M.S. from The University of Chicago and was a research fellow at Harvard University.

**Alan Leshner** is the chief executive officer, emeritus, of the American Association for the Advancement of Science (AAAS), the former executive publisher of the journal *Science*, and the chair of the National Academies of Sciences, Engineering, and Medicine's Committee on the Science of Science Communication: A Research Agenda. Previously, Dr. Leshner was the director of the National Institute on Drug Abuse at the National Institutes of Health. He also served as the deputy director and the acting director of the National Institute of Mental Health and in several roles at the National Science Foundation. Before joining the government, he was a professor of psychology at Bucknell University.

Dr. Leshner is an elected fellow of the AAAS, the American Academy of Arts & Sciences, the National Academy of Public Administration, and many other professional societies. He is a member and served on the governing council of the National Academy of Medicine. He was appointed by President George W. Bush to the National Science Board in 2004, and then reappointed by President Barack Obama in 2011. Dr. Leshner received his Ph.D. and M.S. in physiological psychology from Rutgers University. He has been awarded seven honorary doctor of science degrees.

**Ashley Llorens**, Johns Hopkins Applied Physics Laboratory electrical engineer, leads a double life. His "day job" is managing passive sonar automation projects for the U.S. Navy. On his own time, Llorens has a second career as a musician. Known professionally as SoulStice, Llorens is a lyricist, producer, and internationally acclaimed hip-hop artist.

Growing up in Chicago, he became interested in science performing simple experiments like stripping twist ties and sticking the wires in electrical sockets. That scientific curiosity led him to computers and eventually electrical engineering, his major at the University of Illinois (B.S., 2001; M.S., 2003, in electrical engineering).

**Arthur Lupia** is the Hal. R. Varian Collegiate Professor of Political Science at the University of Michigan. Dr. Lupia studies decision making and learning, and he uses this information to convey complex ideas to diverse audiences and to improve decision making and the communication of scientific facts. He is the former chair of the National Academies of Sciences, Engineering, and Medicine's Roundtable on the Application of Social and Behavioral Science Research and serves on the boards of organizations dedicated to increasing the social value of scientific research, including the Center for Open Science, Climate Central, and the National Academies' Division of Behavioral and Social Sciences and Education. He is an elected member of the American Academy of Arts & Sciences and an elected fellow of the American Association for the Advancement of Science.

Dr. Lupia's articles appear in political science, economics, and law journals, and his editorials are published in leading newspapers. His research has been supported by a wide range of groups, including The

World Bank, The Public Policy Institute of California, the Markle Foundation, and the National Science Foundation. In 2016, Oxford University Press released his latest book, *Uninformed: Why People Know So Little About Politics and What We Can Do About It*. Dr. Lupia earned his M.S. and Ph.D. in social science from the California Institute of Technology.

**Charles Manski** is the Board of Trustees Professor of economics in the Department of Economics and the Institute for Policy Research at Northwestern University. Dr. Manski's research spans econometrics, judgment and decision, and the analysis of social policy. He is the author of six books, including *Public Policy in an Uncertain World* and *Identification for Prediction and Decision*. Dr. Manski has served as director of the Institute for Research on Poverty at the University of Wisconsin–Madison and as editor of the *Journal of Human Resources*.

At the National Academies of Sciences, Engineering, and Medicine, Dr. Manski serves on the Committee on Advancing Social and Behavioral Science Research and Application Within the Weather Enterprise, the Committee on Proactive Policing, and the Report Review Committee. He previously chaired the Committee on Data and Research for Policy on Illegal Drugs.

Dr. Manski received his Ph.D. in economics from the Massachusetts Institute of Technology. He is an elected member of the National Academy of Sciences and a fellow of numerous other organizations, including the American Economic Association, American Statistical Association, American Academy of Arts & Sciences, and The Econometric Society.

**Douglas Massey** is the Henry G. Bryant Professor of Sociology and Public Affairs at the Woodrow Wilson School of Public and International Affairs at Princeton University. Dr. Massey specializes in the sociology of migration and has written on the harmful effects of residential segregation in the United States. He has authored numerous books, most recently *Spheres of Influence: The Social Ecology of Racial and Class Inequality*, co-authored with Stefanie Brodmann; *Climbing Mount Laurel: The Struggle for Affordable Housing and Social Mobility in the American Suburb*, co-authored with Len Albright, Rebecca Casciano, Elizabeth Dickerson, and David Kinsey; and *Brokered Boundaries: Creating Immigrant Identity in Anti-Immigrant Times*, co-authored with Magaly Sánchez.

Dr. Massey is the president of the American Academy of Political and Social Science and the past president of the Population Association of America and of the American Sociological Association. He currently services on the U.S. Bureau of the Census's Scientific Advisory Board. Dr. Massey has served on the National Academy of Sciences Council and on the National Research Council Governing Board. He has won several awards for his books. Dr. Massey received his Ph.D. and M.A. in sociology from Princeton University and his B.A. in sociology, psychology, and Spanish from Western Washington University.

**Marcia McNutt** is a geophysicist and the 22nd president of the National Academy of Sciences. From 2013 to 2016, she was the editor-in-chief of the journal *Science*. Dr. McNutt was director of the U.S. Geological Survey (USGS) from 2009 to 2013, during which time USGS responded to a number of major disasters, including the *Deepwater Horizon* oil spill. For her work to help contain that spill, Dr. McNutt was awarded the U.S. Coast Guard's Meritorious Service Medal. She is a fellow of the American Geophysical Union (AGU), Geological Society of America, American Association for the Advancement of Science, and the International Association of Geodesy. Her honors include membership in the American Philosophical Society and the American Academy of Arts & Sciences. In 1998, Dr. McNutt was awarded the AGU's Macelwane Medal for research accomplishments by a young scientist, and she received the Maurice Ewing Medal in 2007 for her contributions to deep-sea exploration. Dr. McNutt received her B.A. in physics from Colorado College and her Ph.D. in earth sciences from the Scripps Institution of Oceanography.

**M. Granger Morgan** is the Hamerschlag University Professor of Engineering at Carnegie Mellon University, where he is a professor in the Department of Engineering and Public Policy, a professor in the Department of Electrical and Computer Engineering, the co-director of the Center for Climate and Energy Decision Making, and the co-director of the Electricity Industry Center. Dr. Morgan's research interests are centered on policy problems in which technical and scientific issues play a central role, with a particular focus on energy, environmental systems, and climate change and risk analysis. Much of his work has involved the development and demonstration of methods to characterize and treat uncertainty in quantitative policy analysis.

Dr. Morgan received his Ph.D. in applied physics and information science from the University of California, San Diego, his M.S. in astronomy and space science from Cornell University, and his B.A. in physics from Harvard College. He is a member of the National Academy of Sciences and a fellow of the American Association for the Advancement of Science, the Institute of Electrical and Electronics Engineers, and the Society for Risk Analysis.

**Erin Nash** is an advanced doctoral candidate at the Centre for Humanities Engaging Science and Society/Department of Philosophy at Durham University in the United Kingdom. Before returning to academia, Nash had a decade-long public policy career working in both government and nongovernment organizations in Australia, Europe, and Southeast Asia. Nash's doctoral project is a work of practical philosophy, exploring the links among scientific epistemological issues, speech acts about science, personal and political freedom, and democratic policy making. Nash holds an undergraduate degree in science (Hons.) from Monash University in Australia, and a master's in philosophy and public policy from the London School of Economics and Political Science.

**Illah Nourbakhsh** is a professor of robotics; the director of Robotics, Education and Technology Empowerment; and the associate director for robotics faculty at Carnegie Mellon University. Dr. Nourbakhsh's current research projects explore community-based robotics, including educational and social robotics, and ways to use robotic technology to empower individuals and communities. While on leave from Carnegie Mellon in 2004, he served as robotics group lead at NASA's Ames Research Center. He was founder and chief scientist of Blue Pumpkin Software. Dr. Nourbakhsh earned his B.A., M.A., and Ph.D. in computer science from Stanford University and has been a faculty member of Carnegie Mellon since 1997. In 2009, the National Academy of Sciences named him a Kavli fellow. He is also the chief executive officer and chairman of Airviz, Inc., a company dedicated to empowering individuals regarding home air quality. He is a member of the Global Future Council on the Future of AI and Robotics for the World Economic Forum.

**Brendan Nyhan** is a professor in the Department of Government at Dartmouth College. His research, which focuses on misperceptions about politics and health care, has been published in journals, including the *American Journal of Political Science, British Journal of Political Science, Journal of Politics, Medical Care, Pediatrics, Political Analysis, Political Behavior, Political Psychology, Social Networks,* and *Vaccine*. Before coming to Dartmouth, he was a Robert Wood Johnson Foundation Scholar in Health Policy Research at the University of Michigan. Nyhan has also been a contributor to *The New York Times* website *The Upshot* since its launch in 2014. He previously served as a media critic for the *Columbia Journalism Review*; co-edited *Spinsanity*, a non-partisan watchdog of political spin that was syndicated in *Salon* and the *Philadelphia Inquirer*; and co-authored *All the President's Spin*, a *New York Times* best-seller that Amazon.com named one of the 10 best political books of the year in 2004.

**KerryAnn O'Meara** is a professor of higher education and affiliate faculty in women's studies at the University of Maryland, College Park (UMD), where she teaches courses on the academic profession, organizational change in higher education, women in higher education, ranking systems in higher education, and doctoral proseminars. Dr. O'Meara is also the director of the UMD ADVANCE Program for Inclusive Excellence, which began in 2010 as a 5-year, National Science Foundation–funded campus-wide project promoting institutional transformation with respect to the retention and advancement of women faculty. As the program director, she collaborates with colleagues nationally and internationally to shape academic reward system reform, both to support newer forms of scholarship and to reform faculty roles and rewards. The ADVANCE Program aims to improve the faculty work environment and advance gender equity. Prior to UMD, Dr. O'Meara spent 2 years working as a research associate at Harvard University's Project on Faculty Appointments and 6 years on the faculty at the University of Massachusetts Amherst.

Dr. O'Meara received her B.A. in English literature from Loyola University in Maryland, her M.A. in higher education from The Ohio State University, and her Ph.D. in education policy from UMD.

**Matthew Porteus** is an associate professor of pediatrics at Stanford University, where he focuses on hematopoietic stem cell transplantation and pediatric hematology-oncology. Dr. Porteus completed his A.B. degree in history and science at Harvard University and his combined M.D./Ph.D. at Stanford Medical School, with his Ph.D. focused on understanding the molecular basis of mammalian forebrain development. After the completion of his dual-degree program, he was an intern and resident in pediatrics at Boston Children's Hospital and then completed his pediatric hematology-oncology fellowship in the combined Dana-Farber Cancer Institute/Boston Children's Hospital program. He began his studies in developing homologous recombination as a strategy to correct disease-causing mutations in stem cells as definite and curative therapy for children with genetic diseases of the blood. His research program continues to focus on developing genome editing by homologous recombination as curative therapy for children with genetic diseases. Dr. Porteus attends at the Lucile Packard Children's Hospital, where he takes care of pediatric patients undergoing hematopoietic stem cell transplantation.

**Kenneth Prewitt** is the Carnegie Professor of Social Affairs and the vice president for Global Centers at Columbia University's School of International and Public Affairs, where he is also the director of the Scholarly Knowledge Project. Prior to coming to Columbia, Dr. Prewitt was an assistant professor at The University of Chicago from 1965 to 1984, and from 1998 to 2001 he was the director of the U.S. Census Bureau. He has also served as the director of the National Opinion Research Center, the president of the Social Science Research Council, as the senior vice president of The Rockefeller Foundation, and as the dean of the Graduate School at the New School University. He is a Lifetime National Associate of the National Academies of Sciences, Engineering, and Medicine. Dr. Prewitt received an M.A. from Washington University and a Ph.D. in political science from Stanford University. He was a Danforth Fellow at the Harvard Divinity School. He also received honorary degrees from Southern Methodist University.

**David Rousseau** is the vice president and the executive director of health policy media and technology for The Henry J. Kaiser Family Foundation. He oversees the foundation's health policy media programs, including Kaiser Health News and all journalism programs, and directs the foundation's technology and online activities. Rousseau was the director of the foundation's state health facts project and was an associate director of the Kaiser Program on Medicaid and the Uninsured. His work at the Kaiser Program was focused on Medicaid and Children's Health Insurance Program spending, financing, and enrollment, as well as Medicaid service delivery issues, including managed care. He has also managed a range of projects relating to health reform, access to care, and health spending. Rousseau has been a member of the adjunct faculty of The George Washington University's Milken Institute School of Public Health as a lecturer in the Department of Health Policy. His work has appeared in various journals, including *Health Affairs* and the *Journal of the American Medical Association*.

**Dietram Scheufele** is the John E. Ross Professor in Science Communication in the Department of Life Sciences Communication at the University of Wisconsin (UW)–Madison, and Vilas Distinguished Achievement Professor at UW–Madison and in the Morgridge Institute for Research. Since 2013, Dr. Scheufele has also held an honorary professorship at the Dresden University of Technology in Germany. He was the vice chair of the National Academies of Sciences, Engineering, and Medicine's Committee on the Science of Science Communication: A Research Agenda.
Dr. Scheufele is a fellow of the American Association for the Advancement of Science; the International Communication Association; and the Wisconsin Academy of Sciences, Arts & Letters; and a member of the German National Academy of Science and Engineering. He has been a tenured faculty member at Cornell University, a Shorenstein Fellow at Harvard University, and a visiting scholar at the Annenberg Public Policy Center at the University of Pennsylvania. He received an M.A. in journalism and mass communications and a Ph.D. in mass communications with a minor in political science from UW–Madison.

**Susan Scrimshaw** is the co-chair of the Board of Directors for the Nevin Scrimshaw International Nutrition Foundation and the former president of The Sage Colleges in Troy, New York. In addition to the numerous positions Dr. Scrimshaw has held over the course of her 29-year higher education career, she is a medical anthropologist and has published widely on community participatory research methods, addressing health disparities, improving pregnancy outcomes, violence prevention, health literacy, and culturally appropriate delivery of health care. She is a member of the National Academy of Medicine, where she was elected a member of the governing council, and served on the National Academies of Sciences, Engineering, and Medicine's Committee on Science, Engineering, and Public Policy. Additionally, she is the co-chair of the National Academies' Global Forum on Innovation in Health Professional Education. Dr. Scrimshaw is also a fellow of the American Association for the Advancement of Science, the American Anthropological Association, and the Institute of Medicine of Chicago. She is a past president of the board of directors of the U.S.-Mexico Foundation for Science, former chair of the Association of Schools of Public Health, and a past president of the Society for Medical Anthropology. Her honors and awards include the Margaret Mead Award, a Hero of Public Health gold medal awarded by President Vicente Fox of Mexico, the University of Illinois at Chicago Mentor of the Year Award in 2002, and the Chicago Community Clinic Visionary Award in 2005. Dr. Scrimshaw is a graduate of Barnard College and obtained her M.A. and Ph.D. in anthropology from Columbia University.

**Frank Sesno** is an Emmy Award–winning journalist and director of The George Washington University's (GW's) School of Media and Public Affairs. He is also the creator of GW's Planet Forward, where he hosts and facilitates the Planet Forward Salon Series, which focuses on topics such as energy policy, green jobs, and food production. Sesno's journalism career spans more than 25 years, including 21 years at CNN. He served as CNN's DC bureau chief, anchor, and White House correspondent; was the long-running host of CNN's Sunday talk show *Late Editions*; and is now a frequent guest host for CNN's *Reliable Sources*. Sesno has covered a diverse range of subjects from politics and conventions to international summits and climate change. He has interviewed five U.S. presidents and thousands of political, business, and civic leaders—ranging from Hillary Clinton and Israeli Prime Minister Benjamin Netanyahu to Microsoft founder Bill Gates and broadcast legend Walter Cronkite. Sesno received his B.A. from Middlebury College.

**Ahna Skop**, geneticist and artist, is an associate professor in the Department of Genetics at the University of Wisconsin (UW)–Madison and an affiliate faculty member in Life Sciences Communication at the UW–Madison Arts Institute. Dr. Skop's lab seeks to understand the molecular mechanisms that underlie cell division during embryonic development. Some of her artwork can be seen in the main entrance of the Genetics-Biotechnology Center on the UW–Madison campus. Dr. Skop is a winner of the prestigious Presidential Early Career Awards for Scientists and Engineers. In 2008, she was awarded an honorary doctorate of science from the College of St. Benedict, and was named a Remarkable Women in Science from the American Association for the Advancement of Science. In 2015, she was selected as a Kavli Fellow by the National Academy of Sciences. Her science and art have been featured by Apple, NPR, PBS, *Science*, *The Scientist*, *Smithsonian Magazine*, and *USA Today*. Dr. Skop received her Ph.D. from UW–Madison and completed her postdoctoral studies at the University of California, Berkeley.

**Jack Stewart** is a senior writer at *Wired Magazine*, covering the rapidly changing world of transportation. Previously, Stewart was a BBC radio journalist for more than a decade, reporting on science research, technology, and news stories from all over the world. He hosted *Science in Action*—the longest running show on the BBC World Service—and covered breaking news from the Iraq War to the London Underground bombings. Through his current work at *Wired*, Stewart guides a global audience of more than 150 million through the world of transportation research, with a look at how the future of planes, trains, drones, hyperloops, and electric cars will affect us all. Stewart graduated from Brunel University with a degree in mechanical engineering and also attended Sheffield Hallam University, where he studied broadcast journalism.

**Elizabeth Suhay** is an assistant professor of government in the School of Public Affairs at American University. Dr. Suhay specializes in political psychology, political communication, and the intersection of politics and scientific knowledge, mainly within the U.S. context. She has published more than a dozen articles and co-edited, with James Druckman, a recent issue of *The ANNALS of the American Academy of Political and Social Science* titled *The Politics of Science*. She is currently working on a book manuscript on the reciprocal relationship between the American public's political preferences and their causal explanations for socioeconomic inequality. Data collection for this project is being funded by the Russell Sage Foundation. She received her Ph.D. in political science from the University of Michigan.

**Emmanuel Taylor** is a senior electricity consultant at Energetics Incorporated and was formerly an electrical engineer at the U.S. Department of Energy. He possesses a range of professional experience, covering hardware and system design, software development, energy policy, academic research, and technical consulting. Dr. Taylor holds a B.S., M.S., and Ph.D. in electrical and computer engineering from the University of Pittsburgh.

His current work includes strategic planning, technology roadmapping, science communication, and microgrid design. His expertise is in electric power systems, power electronics, and renewable energy.

**Kathleen Tierney** is a professor of sociology at the University of Colorado Boulder, where she served as the director of the Natural Hazards Center from 2003 until 2016. During her career, she studied a wide range of disasters, including earthquakes in Haiti, Japan, and the United States; major hurricanes such as Andrew, Hugo, and Katrina; various technological disasters; and the terrorist attacks of September 11, 2001, in New York City. She is the senior author of *Facing the Unexpected: Emergency Preparedness and Response in the United States*, co-editor of *Emergency Management: Principles and Practice for Local Government*, and is currently completing a book titled *Social Foundations of Risk and Resilience*.

Dr. Tierney has served as a member of several National Academies of Sciences, Engineering, and Medicine committees including the Committee on Disaster Research in the Social Sciences, the Panel on Strategies and Methods for Climate-Related Decision Support, the Panel on Informing Effective Decisions and Actions Related to Climate Change, and the Committee to Advise the U.S. Global Change Research Program. Additionally, she serves on the steering committee of the American Sociological Association's Task Force on Climate Change and on the board of directors of the Earthquake Engineering Research Institute, and she is the co-editor of the *Natural Hazards Review*. Dr. Tierney received the Earthquake Engineering Research Institute's Distinguished Lecturer Award in 2006 and the Fred Buttel Award for Distinguished Contributions from the American Sociological Association's Section on Environment, Technology, and Society in 2012. Dr. Tierney received her M.A. and Ph.D. in sociology from The Ohio State University.

**Doron Weber** is the vice president of programs and the program director at the Alfred P. Sloan Foundation. In this role, Weber helps the president oversee and improve all aspects of the Sloan Foundation's programs and plays a leadership role in the organization's broader philanthropic efforts within the foundation community. For the past 20 years, Weber has run the program for the Sloan Foundation's Public Understanding of Science, Technology & Economics, which uses diverse media—books, radio, television, film, theater, opera, and new media—to bridge the two cultures of science and the humanities and to educate and engage the public. Weber also directs the Sloan Foundation's efforts to promote the

Universal Access to Knowledge through the Digital Information Technology program, which seeks to utilize emerging developments in information technology to make the benefits of human knowledge and human culture safely accessible for people everywhere.

**Laurie Weingart** is the interim provost and the Richard M. and Margaret S. Cyert Professor of Organizational Behavior and Theory at the Tepper School of Business at Carnegie Mellon University. Dr. Weingart's research examines negotiation, conflict, and innovation in teams. Her early research focused on group processes and on social motives and tactical behavior in negotiation. Her more recent research examines cognition, conflict, emotion, and innovation in cross-functional teams. Dr. Weingart has published more than 50 articles and book chapters in the fields of management, social psychology, industrial psychology, and cognitive psychology. Additionally, she has served as the chair of the Conflict Management Division of the Academy of Management in 2001, the president of the International Association for Conflict Management in 2003–2004, and the founding president of the Interdisciplinary Network for Group Research in 2007–2012. She is currently the co-editor of the *Annals of the Academy of Management*. Dr. Weingart earned her Ph.D. in organizational behavior from the Kellogg School of Management at Northwestern University in 1989.